Philip James Rufford

Marine Zoology

Philip James Rufford

Marine Zoology

ISBN/EAN: 9783954272686
Erscheinungsjahr: 2013
Erscheinungsort: Bremen, Deutschland

© *maritimepress in Europäischer Hochschulverlag GmbH & Co. KG, Fahrenheitstr. 1, 28359 Bremen. Alle Rechte beim Verlag und bei den jeweiligen Lizenzgebern.*

www.maritimepress.de | office@maritimepress.de

Bei diesem Titel handelt es sich um den Nachdruck eines historischen, lange vergriffenen Buches. Da elektronische Druckvorlagen für diese Titel nicht existieren, musste auf alte Vorlagen zurückgegriffen werden. Hieraus zwangsläufig resultierende Qualitätsverluste bitten wir zu entschuldigen.

ZOOLOGY

MARINE ZOOLOGY

UPON inquiring into the subject of the marine zoology of the Sussex coast, one is struck with the fact how few systematic and published records exist on that subject.

Natural history societies, more or less organized, appear to have done very little hitherto in recording and publishing the local fauna and flora, and it has devolved upon individuals with sufficient enthusiasm and love of the subject to undertake the task. In the case of the Hastings district, however, one is fortunate in finding a good and solid foundation laid for future investigators in *The Natural History of Hastings and St. Leonards and the Vicinity*, with its three supplements issued at various periods, edited by the Rev. E. N. Bloomfield, M.A., F.E.S., rector of Guestling, and Mr. E. A. Butler, B.A., B.Sc. The former gentleman kindly affords the information that Mr. Butler edited the fauna of the original number and of the first supplement, and that the two later supplements were edited by himself.

As regards the individual workers in the domain of marine zoology who have supplied the data upon which the publication was to some extent founded, may be mentioned the late Dr. Bowerbank, who, as is well known, worked out the sponges, a great proportion of which were obtained from the Diamond Ground off Hastings. The list of sponges in the original number of the Hastings Natural History was apparently compiled by the Rev. Mr. Bloomfield from Bowerbank's *Monographs of the British Spongiadæ*, and afterwards revised by Bowerbank shortly before his death.

Of the Hydrozoa, Mr. Tumanowicz appears to have left a legacy from *Hastings past and present*, and Miss Jelly and Mr. R. Hope, F.Z.S., apparently amplified the list in the original number, the latter also contributing to the second supplement.

For the Vermes Mr. Butler appears to be mainly responsible. The Polyzoa of the original list and most of the first supplement are attributable to Miss Jelly, and those of the later supplement, in the main, to Mr. R. Hope. In the sphere of the Mollusca Mr. A. H. Langdon contributes the list in the original number and in the first and second supplements.

In the ensuing lists of marine animals which appear in this paper, all those species which are quoted from the Hastings Natural History above referred to are distinguished by an asterisk thus (*)

Another source, whence a considerable and substantial addition to the present list of the fauna of this coast has been made, is a publication entitled *The Natural History of Brighton*, by Mrs. Merrifield. Some time has elapsed since the appearance of this book, but there is no reason to suppose that the fauna has since undergone much alteration in character. The whole of the species quoted in the following lists as occurring at Brighton (unless otherwise stated) which have been derived from this little work are distinguished by a dagger mark thus (†)

Only those species which have come within the personal ken of the present writer have received descriptive accounts.

A great proportion of the specimens referred to hereafter have been obtained from the Diamond Ground off Hastings, a considerable area of which consists of sand, but some parts of it are rough ground. Masses of rock, each weighing several hundredweight, are frequently brought up in the trawl. These blocks of stone are the hard, resisting residue left from the demolition of the Wealden rocks, which constitute the coast line between the chalk downs of Eastbourne on the west and those of Folkestone on the east. The softer clays and sandstones being readily disintegrated, leave the hard ironstones to continue a longer existence. Nearer the downs on either hand are found masses of Cherty Green-sand and large flints derived from the Chalk.

A section of the Channel, due south of Hastings, shows a gradual slope attaining to a depth, at fifteen miles from shore, of twenty fathoms, rapidly deepening to thirty fathoms, which depth is maintained for about ten miles; after that there is a slight and gradual rise toward mid-channel.

The Diamond Ground, from the fisherman's point of view, commences at about the twenty-five fathom line, up to which point the ground is of a more or less rough description, and beyond this line it appears to extend for a somewhat unlimited distance. To the west-ward of Beachy Head, or the West Diamond Ground, as it is called, the ground is rough.

Off Rye and Dungeness there is mainly sand, but some six or seven miles from shore there is a deep deposit of mud extending in an easterly and westerly direction, called by the fishermen the Trail, and much frequented by certain kinds of fish. Still further on, rocky ground is met with, known as the East Shoal or Gringer Shoal. From the above short description therefore it will be inferred that the diversified character of the sea-bottom off Hastings affords a suit-able ground for the varied and rich fauna which it possesses.

In the preparation of the present article, recourse has been had amongst other works to the following; and in the case of the Hydroid Zoophytes and the Polyzoa the nomenclature of Hincks has been adopted, whilst in the Mollusca the list of the Conchological Society has been followed.

MARINE ZOOLOGY

FORAMINIFERA.—*On the Recent Foraminifera of Great Britain*, W. C. Williamson, Ray Society ; 'Report on the Foraminifera' (*Challenger Expedition*), H. B. Brady.

SPONGES.—Article upon 'Sponges' in the *Encyclopedia Britannica*, W. J. Sollas ; *A Monograph of the British Spongiadæ*, J. S. Bowerbank, Ray Society.

CŒLENTERA.—*A History of the British Hydroid Zoophytes*, Thomas Hincks ; *A History of the British Sea Anemones and Corals*, P. H. Gosse ; *Die Ctenophoren des Golfes von Neapel und der Angrenzenden Meeres-Abschnitte*, Dr. Carl Chun.

VERMES.—*A Catalogue of the British Non-parasitic Worms in the Collection of the British Museum*, George Johnston.

POLYZOA.—*A History of the British Marine Polyzoa*, Thomas Hincks.

ECHINODERMATA.—*A Catalogue of the British Echinoderms in the British Museum* (*Nat. Hist.*), F. Jeffrey Bell.

MOLLUSCA AND TUNICATA.—*A History of the British Mollusca*, Forbes and Hanley ; *A Monograph of the British Nudibranchiate Mollusca*, Alder and Hancock, Ray Society ; 'Report on the Tunicata' (*Challenger Expedition*), William A. Herdman.

GENERAL.—*A Manual of Marine Zoology for the British Isles*, P. H. Gosse.

A summary of the species detailed in the following lists will be found as follows :—

Protozoa	7 species
Porifera	58 ,,
Cœlentera	77 ,,
Vermes	31 ,,
Polyzoa	109 ,,
Echinoderma	25 ,,
Brachiopoda	1 ,,
Mollusca	192 ,,
Tunicata	14 ,,
Total	514 ,,

PROTOZOA

RHIZOPODA

FORAMINIFERA

1. *Nonionina crassula*, Walker.

Shell nautiloid, umbilicated, having eight or nine segments. Colour, a dull crystalline. Upon corallines, polyzoa, etc. Not uncommon. Hastings.

2. *Polystomella crispa*, Linnæus.*

Not uncommon. Hastings.

3. *Polystomella umbilicatula*, Walker.

Common. Hastings.

4. *Rotalina beccarii*, Linnæus.*

Hastings.

5. *Miliolina seminulum*, Linnæus.

Shell ovate, milkwhite, with two segments to a volution ; aperture alternately facing opposite ways. Common upon corallines. Hastings.

6. *Halyphysema tumanowiczii*, Bowerbank.

A species occurring upon seaweed and corallines. It is somewhat club-shaped, curved, tapering downwards to an expanded disc-like base. The test, more particularly the distal end, bristles with attached sponge spicules, arranged radially ; grains of sand are also attached.

Under the above title this species appears described by Bowerbank among the Porifera in his *Monograph of British Spongiadæ*. It has since however been definitely referred to the Foraminifera (fam. Astrorhizidæ). Somewhat common. Hastings.

INFUSORIA

FLAGELLATA

7. *Noctiluca miliaris*.

The phosphorescence of the sea water is often due to these minute globular creatures. They are about the size of a pin's head, and emit the light more particularly when the water is agitated. In countless numbers. Hastings.

PORIFERA

CALCISPONGIÆ

1. *Leucosolenia (Ascon) botryoides*, Bowerbank.

This sponge occurs in little groups upon

corallines. When examined under the microscope it seems hardly to resemble bunches of grapes as the specific name would imply, but rather the fingers and shallow palm of an irregular glove, more particularly if we may suppose the fingers to give off other fingers, the ends being open to represent the oscula. The simple sponge is cylindrical, slightly swollen towards the end, with a wide paragaster and osculum. From moderately deep water. Rare. Hastings.

2. *Leucosolenia (Ascon) lacunosa*, Bowerbank.

Sponge white, fig-shaped, compressed, with a short stem, and with large oval orifices occurring over the whole surface. Bowerbank describes this as a very rare sponge. The form of the Hastings specimen is somewhat intermediate between Bowerbank's two figures.

The dimensions are : Height of stem, 1·5 mm. ; body, 6·5 mm. ; greatest width, 6 mm.; lesser diameter, 3 mm.

A single specimen only taken ; growing upon an Eudendrium. From moderately shallow water. Very rare. Hastings.

3. *Grantia (Sycon) ciliata*, Fleming.

This is a simple sponge consisting of a small white cylinder, the surface bristling with defensive spicules which project in minute tufts over the blind ends of the radial canals. Around the terminal osculum there is a fringe of spicules which suggest the specific name. Specimens taken from the rocks at low water are fully double the size of those found upon corallines, etc., from deeper water, and are of a more attenuated form, and have the oscular spicules more conspicuous. Not uncommon, but small. Hastings.

4. *Grantia (Sycon) compressa*, Fleming.

This species forms little grey or tan coloured sacks, cylindrical to ovate in form, generally with a single osculum. Hastings specimens do not exceed 1½ inches in height. Upon rocks at low tide ; not noted from deeper water. Not uncommon. Hastings.

PLETHOSPONGIÆ

RHAGON

5. *Ecionemia ponderosa*, Bowerbank.*
Hastings.

6. *Ciocalypta penicillus*, Bowerbank.*
Hastings.

7. *Ciocalypta*, sp.

A specimen taken from the rocks at low water. Mr. R. Kirkpatrick, of the British

Museum of Natural History, who very kindly examined the specimen, expressed the opinion that it might possibly be a strongly marked variety in the young condition of *C. penicillus*, though it showed much divergence from the typical adult sponge. The specimen is in the South Kensington Museum. Rare. Hastings.

8. *Tethya lyncurium*, Johnston.

Sponge hemispherical, about ¾ inch in diameter ; surface warty-looking ; colour when fresh, orange. Upon rock from the Diamond Ground. Somewhat rare. Hastings.

9. *Raspailia cristata*, Montague.
Dictyocylindrus ramosus, Bowerbank.

Sponge 4 or 5 inches in height ; brown, branching, and hispid with defensive spicules. The shoots before branching are often palmate. From moderately shallow water. Common. Hastings.

10. *Raspailia ramosa*, Montague.

Not to be confounded with *Dictyocylindrus ramosus* of Bowerbank, the present species branching in the same plane, the branches being somewhat flattened. From the Diamond Ground. Rather rare. Hastings.

11. *Dictyocylindrus hispidus*, Bowerbank.

Sponge, light brown, hispid, dichotomously branching, the branches being in the same plane and curving inwards towards the ends. A fine specimen measures 12 inches in height. From the Diamond Ground. Somewhat rare. Hastings.

12. *Dictyocylindrus fascicularis*, Bowerbank.

Specimens a little over 5 inches in height, of a pale yellow colour, dichotomously branching, the branches being slender and curving inwards towards the upper parts. A dried specimen might be readily mistaken for a dead twig. From the Diamond Ground. Somewhat rare. Hastings.

13. *Dictyocylindrus radiosus*, Bowerbank.

Specimen 3½ inches in height, resembling to some extent *D. hispida*. The branching is dichotomous, and in the same plane, the branches being rather flattened and hispid, and instead of curving inwards, as in the last mentioned species, they expand in a radiating manner. From moderately deep water. Rare. Hastings.

14. *Dictyocylindrus aculeatus* (?), Bowerbank.

The Hastings specimens are in the dried condition, and present little grey feathery tufts about 2 inches in height, closely resembling the illustration of the Northumberland specimen figured by Bowerbank, but of twice the height. The writer learns that similar forms to the present in the British Museum are labelled by Bowerbank *D. ramosus*, Bk., but that Carter labels similar specimens *Raspailia aculeata* (? Bk. sp.). The present specimens were cast up on the beach after a storm, and similar forms have not been since noticed. Hastings.

15. *Microciona fictitia*, Bowerbank.*
Hastings.

16. *Microciona fallax*, Bowerbank.*
Hastings.

17. *Microciona plumosa*, Bowerbank.*
Hastings.

18. *Microciona atrasanquinea*, Bowerbank.*
Hastings.

19. *Microciona spinarcha*, Carter.*
Hastings.

20. *Hymeraphia stellifera*, Bowerbank.*
Hastings.

21. *Raphiodesma sordida*, Bowerbank.*
Hastings.

22. *Hymeniacidon lactea*, Bowerbank.*
Hastings.

23. *Hymeniacidon (Suberites) denuencula*, Olivi.
Sponge white, smooth, rounded and compressed, very solid, no pores or oscula apparent. Size 2 inches by 3 inches by ¾ inch in thickness. The specimen has been apparently attached to a rock. From moderately shallow water. Rare. Hastings.

24. *Hymeniacidon caruncula*, Bowerbank.*
Hastings.

25. *Hymeniacidon mammeata*, Bowerbank.*
Hastings.

26. *Hymeniacidon crustula*, Bowerbank.
Sponge somewhat cylindrical with rounded ends, smooth, orange coloured. Length, 4 inches by 1¼ inches in diameter. The specimen envelops the stems of a coralline. From moderately shallow water; rather rare. Hastings.

27. *Hymeniacidon suberea*, Bowerbank.*
Hastings.

28. *Hymeniacidon (Clione) celata*, Bowerbank.*
The species is dark brown in colour, and is found boring into shells and rock. It frequently bores through the valves of the scallop, obliging the animal to barricade itself within by fresh excretions of shelly matter. From the Diamond Ground; common. Hastings.

29. *Hymeniacidon pannicea*, Johnston.
Sponge green, drying almost white. It incrusts the rocks at low water in considerable masses, often forming ridges of volcano-like vents. Finer specimens are obtained from deeper water, often upon tubularian stems and other hydroids. The deep water specimens are not so characterized by the ridges of vents, but are smoother. Very common. Hastings.

30. *Hymeniacidon glabra*, Bowerbank.*
Hastings.

31. *Hymeniacidon distorta*, Bowerbank.
Sponge rather low and branching; colour, grey-brown. Before branching the shoots are palmate, giving off others in a plane at a right angle. Texture of sponge somewhat velvet-like. Scattered here and there over the surface are stellate oscula. From moderately shallow water; rather rare. Hastings.

32. *Hymeniacidon corrugata*, Bowerbank.*
Hastings.

33. *Hymeniacidon incrustans*, Johnston.*
Hastings.

34. *Hymeniacidon irregularis*, Bowerbank.*
Hastings.

35. *Hymeniacidon nigricans*, Bowerbank.*
Hastings.

36. *Hymeniacidon pattersoni*, Bowerbank.*
Hastings.

37. *Hymeniacidon ingalli*, Bowerbank.*
Hastings.

38. *Hymeniacidon farinaria*, Bowerbank.
Forming a pale buff-coloured coating of fine texture upon shell of *Pecten opercularis*, from the Diamond Ground off Hastings.

39. *Isodictya cinerea*, Bowerbank.*
Hastings.

40. *Isodictya indistincta*, Bowerbank.*
Hastings.

41. *Isodictya pallida*, Bowerbank.*
Hastings.

42. *Isodictya hyndmani*, Bowerbank.*
Hastings.

43. *Isodictya mammeata*, Bowerbank.*
Hastings.

44. *Isodictya simulans*, Johnston.

Sponge rather low and straggling, branching, the branches cylindrical or slightly compressed and anastomosing. Oscula distinct and upon one side only of the branches. Texture, fine; colour, ash-grey to brown. From moderately shallow water; somewhat rare. Hastings.

45. *Isodictya dichotoma*, Bowerbank.*
Hastings.

46. *Isodictya fucorum*, Bowerbank.*
Hastings.

47. *Isodictya rugosa*, Bowerbank.*
Hastings.

48. *Isodictya obscura*, Bowerbank.*
Hastings.

49. *Desmacidon fruticosa*, Bowerbank.

Sponge extensive, low, spreading, coarse in texture, grey, and giving off short wide funnel-like branches with wide terminal orifice, which also extends partly down the side. Growing rather insecurely upon two or three stones. From the Diamond Ground; somewhat rare. Hastings.

50. *Desmacidon ægagropila*, Bowerbank.*
Hastings.

51. *Desmacidon copiosa*, Bowerbank.*
Hastings.

52. *Desmacidon rotalis*, Bowerbank.*
Hastings.

53. *Raphyrus griffithsii*, Bowerbank.

Sponge bark brown, forming rounded masses upon stones, etc., the whole surface being closely pitted. From moderately deep water; somewhat rare. Hastings.

54. *Chalina oculata*, Bowerbank.

Sponge with a pedicel. Branches close and compact, and given off somewhat in the same plane. In a general way the oscula are arranged upon two opposite sides of the branches, but this order is by no means constant. A fine specimen measures 12 inches high. From the Diamond Ground; common. Hastings.

55. *Chalina montaguii*, Bowerbank.*
Hastings.

56. *Chalina gracilenta*, Bowerbank.*
Hastings.

57. *Dysidea fragilis*, Bowerbank.

Sponge forming somewhat shapeless or lobed masses, growing upon rock, etc. The fibres are cored with sand grains, and the sponge when dried is extremely fragile. Spicules are practically absent in this genus. Trawled in moderately shallow water; not uncommon. Hastings.

58. *Dysidea coriacea*, Bowerbank.*
Hastings.

CŒLENTERA
HYDROZOA
HYDROIDA
ATHECATA

CLAVIDÆ

1. *Clava multicornis*, Forskål.

Polypite naked, spindle-shaped, semi-opaque white; tentacles many and long, distributed irregularly over the body; gonophores round and borne below the tentacles. Upon shells and under stones at low water. Colonies small; somewhat rare. Hastings.

HYDRACTINIIDÆ

2. *Hydractinia echinata*, Fleming.

Colonies incrusting various shells occupied by the hermit crab, more particularly those of the whelk, natica and nassa; also noted upon claw of lobster.

This zoophyte is peculiar for the specialization of its members and the form of its polypary. The alimentary polypite is naked, columnar, tapering downwards, and with a single circlet of tentacles. The pink gonophores are borne on modified polypites, giving to the colony when very prolific a delicate rose colour. There are two other kinds of Zoöids, one forming coils and ostensibly a modified polypite and the other long and very contractile with bilobed 'head.' The functions of these two members are problematical. Sections of the crust show superimposed reticulating galleries formed of chitine and traversed by cœnosarcal threads. Comparison may be made with advantage with sections of the polyparies of *Coppinia arcta*, *Antennularia ramosa* and other species. Very common upon the shore in warm weather; upon the approach however of cold weather the crab retires to deeper water. Hastings.

CORYNIDÆ

3. *Coryne van-benedenii*, Hincks.

Polypite small, club-shaped, with knobbed tentacles, the latter being dispersed over the body and numbering about nineteen. Colour semi-translucent white with opaque white dots. The polypary expands over the base of the polypite, below which it is plain or very slightly undulatory, and throughout the lower three-fourths of the stem it is lightly annulated. The polypary is colourless or of a faint buff shade. Upon *Tubularia indivisa*. Hastings.

4. *Coryne vaginata*, Hincks.

Zoophite 3 inches high, in form resembling a spruce fir tree; polypite naked, rose-coloured, with tentacles scattered as in *Clava*, but short and knobbed at the ends; gonophores oval and borne amongst the tentacles. Common in rock pools, often densely covered with confervæ. Hastings.

EUDENDRIIDÆ

5. *Eudendrium rameum*, Pallas.

Polypite naked, tentacles forming a ring. This zoophyte has been aptly compared to a stunted and weather-beaten tree. The stem and main branches are compound, and when covered with the round and orange-coloured gonophores the colony might be likened to a shrub laden with berries. The gonophores are generally borne upon the cænosarc. Common in the trawl from the Diamond Ground and moderately deep water. Hastings.

6. *Eudendrium ramosum*, Linnæus.

This species suggests a collection of branching twigs. The stem is formed of a single tube, both it and the branches being of a straight and straggling character and of a glossy brown colour. No gonophores observed. Common upon scallops and rock from deep water. Hastings.

7. *Perigonimus repens*, Wright.

Polypite with a single circle of tentacles distant from the mouth; polypary rather coarse, of a red-brown colour, and expanded over the base of polypite to form a rough cup. Taken upon shell of *Nucula nucleus*, in association with *Lovenella clausa*. Coralline zone; rare. Hastings.

8. *Garveia nutans*, Wright.

This species requires more than a passing notice. The only localities given for it by Hincks are Inchgarvie, Firth of Forth, and Shetland. It has been taken some three or four times off shore at Hastings, upon all occasions climbing over Hydrallmania and throwing up short branches, and not as figured by Hincks from northern specimens with erect and compound stem. The polypites have a single circle of tentacles and there is a gradual expansion of the polypary over the base of the polypite. Branches flexuous. The polypites themselves are conspicuous by their colour, which is orange or carrot colour, and which also extends to the cænosarc. The gonophores are likewise orange coloured and are given off from the creeping stolon, emerging from an expansion of the polypary. The Hastings specimens agree fairly well with Hincks' description, but differ in the matter of the compound and erect stem, and in the fact that the polypites were not noticed to nod, from which peculiarity the species takes its name. Somewhat rare. Hastings.

TUBULARIIDÆ

9. *Tubularia indivisa*, Linnæus.

Polypite naked with two crowns of tentacles, the one oral and the other aboral, or midway down the body. Among the latter are borne the gonophores in grape-like bunches. The empty polyparium tubes much resemble tufts of stubble. Not a shore species; the finest specimens are obtained from moderately deep water. Common off Hastings.

10. *Tubularia larynx*, Ellis and Solander.

Except as regards size and habitat there seems little to distinguish this species from *T. coronata*. The Hastings specimens are very faintly annulated, and little or no branching can be detected. The polypites are naked, transparent, very finely spangled with opaque white; gonophores round to oval and borne upon short branched peduncles. The gonophores and manubrium are rose-coloured, and the former have from not any to four tubercles at their distal ends. The gonozooid or extrusion is oval, constricted at the basal end, with twelve long ab-oral tentacles, sometimes less, clubbed at the ends and alternately raised and lowered, by which means this star-like creature stalks about, as upon stilts. At the oral end there are invariably four short thick tentacles, curved inwards. The creature having no bell seems awkward and sluggish, and is apparently intermediate between the fixed and freed forms. This species only visits Hastings in occasional years. In 1897 it appeared in great profusion upon rocks and stones from mid to low tide. Hastings.

THECAPHORA

CAMPANULARIIDÆ

11. *Clytia johnstoni,* Alder.

Colonies trailing over most objects, sea-weed, stones, shells, wood, etc., throwing up delicate partly-ringed stems terminating in calycles with dentate margins. The capsules are annulated and nearly always borne upon the stolon ; gonozoöid medusiform, minutely spotted with opaque white. Very common, ranging from the beach to moderately deep water. Hastings.

12. *Obelia geniculata,* Linnæus.

This very common little species throws up a zig-zag stem from a trailing stolon, giving off at each bend a short branch ending in a plain-rimmed calycle. Capsules borne in the axils. The gonozoöids of *Obelia* have the peculiar habit of often turning the swimming-bell inside out. Common upon weed, stones, shells, etc., upon the beach, and in deeper water. Hastings.

13. *Obelia gelatinosa,* Pallas.

A very beautiful zoophyte suggestive of a young and graceful birch tree. The stem is compound and the branches are usually given off in regular whorls. The calycles are said by Hincks to be dentate, but they are very difficult to define under the microscope, the margin usually appearing folded inwards. The capsules are deep and vase-like and are formed in the axils. A large and common species often growing in very exposed positions on the shore. From imperfect specimens preserved the impression is gained that this species may also occur with simple stolonic stem, over-running other zoophyte stems. Hastings.

14. *Obelia longissima,* Pallas.

A species sometimes over a foot in length, branching and tapering gradually to the summit. The calycles are squarely dentate ; capsules a little deeper than wide. Amongst the trawlers' rubbish it may be readily mistaken for a tangle of hair. Common in the trawl from deep water. Hastings.

15. *Obelia dichotoma,* Linnæus.*
Hastings.

16. *Campanularia integra,* McGillivray.*
Hastings.

17. *Campanularia verticillata,* Linnæus.
Stem and main branches compound. Around the axis are given off simple, partly-ringed branches, rather long and of equal length terminating in dentate calycles. The capsules are long and narrow-necked, and occur on the compound parts of the axis. Not uncommon in the trawl from moderately deep water. Hastings.

18. *Campanularia flexuosa,* Hincks.

The notes and sketches at hand of this species only allow of the remarks that the calycles have a plain margin and are borne upon rather long and well-ringed foot-stalks, and that the capsules are an elongate oval in form. Hastings.

19. *Campanularia neglecta,* Alder.*
Hastings.

20. *Lovénella clausa,* Lovén.

A minute species throwing up long slender stems ringed at the top, undulating elsewhere, with deep elegant calycles of which the scalloped margins are prolonged into pointed segments which meet overhead, closing the aperture. The chitine appears to be of some thickness at the bottom of the calycle, gradually thinning out towards the top ; polypite with from twelve to fourteen tentacles ; no capsules observed. A single specimen associated with *Perigonimus repens* upon *Nucula nucleus.* Coralline zone. Hastings.

21. *Gonothyrea gracilis,* Sars.

This zoophyte at first glance with the hand-glass may be mistaken for *Clytia johnstoni,* but the calycles are much deeper, the teeth of the margin longer and sharper, and inclining inwards rather than outwards. The stem just below the calycle has four or five rings and again at the base is ringed. Branches bearing a terminal polypite are given off at about two-thirds of the distance up the stems. Upon *Tubularia indivisa* ; rare. Hastings.

CAMPANULINIDÆ

22. *Opercularella lacerata,* Johnston.

Zoophyte of very slender habit. It occurs wound like fine thread around the polyzoan *Anguinella palmata,* throwing up short branching stems much annulated. The calycles on short ringed footstalks have the plain margin cut into segments which meet over the centre, forming an operculum. The polypite stretches out of its calycle fully to the extent of the length of the calycle. The species also occurs upon sponges ; not general. Hastings.

LAFOËIDÆ

23. *Lafoëa dumosa*, Fleming.

This species occurs in two forms, either with simple stem over-running other coral-lines and giving off, without footstalks, tubular calycles narrowed and slightly twisted at the base; or it is found with compound branching stem giving off around the axis the closely arranged calycles, the whole suggesting perhaps a very prickly bramble in miniature. Both forms common in the coralline zone. Hastings.

24. *Lafoëa pocillum*, Hincks.

Upon *Diphasia rosacea*, *Eudendrium rameum*, etc. The creeping stem gives off short, ringed peduncles with tubular but shapeable calycles, some of which appear to approach to the more tubular form *pygmæa*. In the Hastings specimens the peduncle has from four to six rings, whereas Hincks gives from five to eight for this species, and two or three to Alder's species *pygmæa*. Somewhat rare. Hastings.

25. *Lafoëa pygmæa* (?), Alder MS.

Some years ago this species was recorded from Hastings, but in the absence of notes and specimens mislaid, a query is here appended.

26. *Calycelia syringa*, Linnæus.

Over-running the polyzoan *Anguinella palmata* together with *Opercularella lacerata*, already noticed. The calycles are borne upon short, three-ringed footstalks given off from the creeping unringed stem. They rather resemble those of *O. lacerata*, but are longer and not so swollen in the middle. Some of the smaller calycles which have the operculum introverted, and so not seen, bear a resemblance also to the calycles of *Lafoëa pocillum* and *pygmæa*. Common upon *Anguinella* at low tide. Hastings.

27. *Filellum serpens*, Hassall.

Stem nearly always creeping over other hydroids, but in one instance upon a scallop shell. It gives off ovate tubular calycles without footstalk, the lower half being adnate, and the upper half curved upwards, showing a slightly trumpet-shaped aperture. Calycles transversely lined.

There is a remarkable form in which apparently this species occurs, not mentioned by another author, so far as the writer is aware, and which merits notice. Upon old shells covered with incrusting polyzoa, the zoœcia of the latter will often be found to contain hydroid calycles peeping out of the apertures and bearing nearest resemblance to the present species. The calycles are always black and glassy, possibly discoloured by sulphuretted hydrogen; sometimes they are long and tubular, at others ovate in the lower, and tubular in the upper half, and always with very trumpet-shaped apertures. There is generally one calycle in each zoœcium, but occasionally there are two. On dissolving the zoœcia in acid, only imperfect calycles are obtained, showing no connection with a stem. It is possible that these may be the primary zoœcia of the present species which are prevented from freely budding by reason of their limited surroundings. The type form is common, and the other form described is not uncommon. From deep water. Hastings.

COPPINIIDÆ

28. *Coppinia arcta*, Dalyell.

A peculiar zoophyte, usually found surrounding in short masses the stem of *Hydrallmania*. A cross-section of the dry polypary shows a chitinous layer enveloping the stem, tunnelled with passages, one above the other. From these passages arise, at a little distance apart, tubular calycles bent in the upper portion at about a right angle. The calycles at half their height are cemented together by a floor of chitine. In the intervening spaces of this floor are seen slightly-tubular orifices, apparently subserving the escape of the planules. Not uncommon. Hastings.

HALECIIDÆ

29. *Halecium halicinum*, Linnæus.

Rather a coarse looking zoophyte. Stem and main branches compound; branches given off pinnately; the footstalks bearing the calycles are telescopic in appearance, the latter resembling in shape a drinking-tumbler. It is important to note in the female gonophores of this genus, as Hincks has pointed out, that the gonozoöid-bearing polypites are not atrophied as in all the rest of the *Thecaphora*, but are perfectly recognizable polypites, protruding from one side of the capsule. From moderate to deep water; common. Hastings.

30. *Halecium beanii*, Johnston.

A species of much more delicate and flexible habit than the last. In the female gonophore there is a lobe which projects considerably in front of the aperture. From deep or moderately deep water; not uncommon. Hastings.

SERTULARIIDÆ

31. *Sertularella polyzonias*, Linnæus.

A little, straggling species with stem and branches of the same thickness throughout. The calycles are somewhat oval and arranged alternately on either side of the axis into which they appear to be sunk. The capsules are large and wrinkled. Often found growing upon annelid tubes, flustra, etc. Common in the trawl. Hastings.

32. *Sertularella gayi*, Lamouroux.

This species resembles in type *S. polyzonias*, but it is larger and of stouter build. The stem is compound with calycles alternate, short, stout and turned well outward, and having four slight denticles to the margin. The polypary is brown, the margins of the calycles appearing somewhat lighter. Not uncommon in the trawl. Hastings.

33. *Sertularella rugosa*, Linnæus.

A small species often over-running flustra, and throwing up short branches with clusters of alternate wrinkled and oval calycles. The capsules resemble the calycles, but are much larger and contracted at the base. Common. Hastings.

34. *Sertularella tenella*, Alder.

Hastings specimens appear to link together as nearly as possible Alder's species *tenella* and Hincks' species *fusiformis*. The branches are about ¼ inch in height, arising from the stem creeping over flustra. Calycles smooth, intermediate in slenderness, aperture with four denticles. The stem is bent at a right angle immediately above each calycle in a strongly zigzag manner. Capsules large, ringed and with four denticles. These points agree therefore with *tenella*, except in the calycles being smooth, not quite so slender, and in the capsules being toothed, in which respects the specimens resemble *fusiformis*. Hastings.

35. *Diphasia rosacea*, Linnæus.

A very delicate and graceful species, the laterally branched stems being flexible and plume-like. The calycles are tubular, bilateral, opposite, and bent straight outwards. The female capsule is thrown into vertical folds producing at the top a crown of spines of which two, one on either side, project, the others being curved over the centre. Habitat, upon other zoophyte stems, sponges, etc., from moderately deep water. Not uncommon. Hastings.

36. *Diphasia attenuata*, Hincks.

A species very like *rosacea* and difficult to determine in the absence of the capsules, which are certainly the best specific guide. In this species the height of the stem joint above the offshoot of the calycles below is not so great as in *rosacea*, and the stem between each pair of calycles is not so attenuated. The calycles are also a trifle longer and narrower. The male capsule has a crown of spines directed horizontally outwards, and one central and vertical spine. Very common from the coralline zone, and from moderately deep water. Hastings.

37. ? *Diphasia fallax*, Johnston.

The species is recorded with a query in the Natural History of Hastings before quoted. Although specimens are not at hand, the record appears well founded. Hastings.

38. *Sertularia pumila*, Linnæus.

This hydroid covers densely the bladder-wrack at low tide. The stem is only about ½ inch in height, and little branched. The calycles are tubular, short, bent outwards and arranged in pairs, oppositely. Capsules ovate. Very common. Hastings.

39. *Sertularia gracilis*, Hassall.

Of very similar growth to the last species but smaller, denser, and altogether more refined. Erect stems, not observed to branch. The calycle margin is thrown into two sharp points. Capsules ovate. This species over-runs other hydroid stems. Not very common. Hastings.

40. *Sertularia operculata*, Linnæus.

This is a rich and luxuriant species and has been termed 'seahair.' It affects mussel shells and *Laminarian* stems. The stems are long, fine, wavy, branching and of equal thickness throughout. The colour might be almost called a dull golden. The calycles are arranged in pairs, oppositely, and the margins of the apertures are thrown into sharp points. Capsules balloon-shaped. Occasionally colonies of this zoophyte might almost be said to rival in the number of its members the population of London. Very common from moderate to deep water. Hastings.

41. *Sertularia filicula*, Ellis and Solander.*

Hastings.

42. *Sertularia abietina*, Linnæus.

The erect stems are about 6 inches in

height, pinnately branched, stout, and with calycles lateral and opposite, to alternate. They are ovately tubular and bent slightly outwards. The capsules are oval and wrinkled. Common upon scallop shells, etc., from moderately deep water. Hastings.

43. *Sertularia argentea*, Ellis and Solander.

Stems of considerable length, gyratory, giving off around the axis short branches in a palmate manner. Calycles sub-opposite ; apertures sharply pointed. Capsules shield-shaped. From moderate to deep water ; not uncommon. Hastings.

44. *Sertularia cupressina*, Linnæus.

Stems very long, branches short and palmate, the zoophyte as a whole tapering to a point in a somewhat snake-like manner. The calycles are sub-opposite, diverge very little from the stem and have sharply-pointed margins. Capsules narrowly shield-shaped. Hastings specimens are rather inferior in size and condition. Common from moderate to deep water. Hastings.

45. *Hydrallmania falcata*, Linnæus.

Stems long and gyratory, giving off around the axis pinnate branches. The calycles are ovately tubular, borne crowded upon the upper sides of the branches, and almost in the same straight line. Their apertures are turned alternately to the right and left. Capsules ovate ; very common from the coralline zone. Hastings.

PLUMULARIIDÆ

46. *Antennularia antennina*, Linnæus.

Stems simple, long and straight, from which are given off radiately at frequent nodes along the axis, short delicate sprays of equal length. The calycles which are cup-like are borne in a single line upon the upper sides of these sprays, and with them are associated the peculiar organs called nematophores. The capsules are ovate. Hincks gives 8 or 10 inches as the height of this species, but the writer has obtained it 18 inches in length from the Diamond Ground, where it is common. This zoophyte is much frequented by the Nudibranch molluscs *Doto coronata* and *D. pinnatifida*, which attach their egg-bands to its stem. Hastings.

47. *Antennularia ramosa*, Lamarck.

The most striking feature of this species is that it branches and rebranches. The stem is compound, a cross-section of it showing a large central tube with many minor ones, varying in size and overlying one another, running parallel with it, the whole being welded together. The tubes communicate one with the other, thus indicating the continuity of the cœnosarc. The calycles and nematophores closely resemble those of the last species, as do also the capsules, but the latter taper towards the base and are curved. Common upon scallops and rock from the Diamond Ground. Hastings.

48. *Aglaophenia pluma*, Linnæus.

This species envelops the stem of *Halidrys siliquosa* in a loose stolonic mesh, giving off beautiful plume-like branches with irregularly toothed calycles arranged in single line upon the upper surfaces of the pinnæ. Associated with the calycles are three nematophores, two lateral and one median. The capsules are ribbed, the ribs being armed with nematophores. Plentiful upon the beach, after rough weather. Hastings.

49. *Plumularia pinnata*, Linnæus.

A very delicate and beautiful species growing in tufts of plume-like stems. The calycles are shallow and cup-like, and arranged singly upon the upper sides of the pinnæ. There are two nematophores, one above and one below each calycle, and one generally situated in the axils of the pinnæ. The gonophores are conspicuously and closely set upon each side of the stem. Hincks observes that the calycles are only separated by a single joint. This does not always appear to hold good with Hastings shore forms, in which there are sometimes two joints. The form from deeper water is much larger but not of frequent occurrence at Hastings, where the shore form is always in profusion on rocks, stones, shells, sponges, etc., at low water. Hastings.

50. *Plumularia setacea*, Ellis.

A most delicate species, almost escaping detection. Readily distinguished from *P. pinnata* by the long drawn out, narrow-necked capsules, when present, or by the difference in the character and number of the nematophores. Taken upon *Antennularia* from deep water ; rare. Hastings.

51. *Plumularia obliqua*, Saunders.*
Hastings.

52. *Plumularia similis*, Hincks.*
Hastings.

SIPHONOPHORA

PHYSOPHORIDÆ

53. *Physalia*, sp.

This record is made for Brighton upon the authority of Mr. W. Wells, superintendent of the Aquarium, Brighton.

LUCERNARIDA

ACRASPEDA

54. *Aurelia aurita.*

In this common jellyfish the umbrella is large and transparent ; the radial canals are of a delicate pale mauve colour, and the tentacles around the margin are many and short. Conspicuous through the umbrella are the opaque-white gonads in quarters. Oral arms short. Common. Hastings.

55. *Chrysaora cyclonota.*

The upper surface of the umbrella is marked around the centre with a brown circular ring, a short distance from which arise brown, V-shaped rays extending to a little distance short of the margin, the surface generally being finely speckled with brown. The marginal lappets are also of a dark brown. Intermediate in position between these are long streaming tentacles. Oral arms long and frilled. Common. Hastings.

56. *Cyanæa lamarckii.*

Looked at from above, the inner surface of the umbrella appears of a pale heliotrope colour, slightly marbled, and around the centre and not far from the margin there is a circular band or coronet of some depth, of a dark heliotrope colour, and sending off rays to the marginal lobes ; these are large, and veined with the branching canals. The tentacles are collected together in knots between the lobes. The surface bordering upon these is strongly cancelled with muscular tissue. This species grows to a large size. Common. Hastings.

ACTINOZOA

ZOANTHARIA

ACTINIARIA

SAGARTIIDÆ

57. *Actinoloba dianthus*, Ellis.

· The disc of this anemone is thrown out into plume-like marginal lobes, covered and fringed with rather small and short tentacles. The column is tall and smooth. The colours are of the most delicate shades, running through every grade of white, pink, red, yellow, salmon, orange, grey

and brown. It is obtained from the Diamond Ground, and may occasionally be met with upon the shore at low tide, but specimens so found have probably been thrown overboard by fishermen. Common. Hastings.

58. *Sagartia bellis*, Ellis and Solander.*
Hastings.

59. *Sagartia miniata*, Gosse.

Animal dark red, as broad as high. Margin of disc thrown into unequal, ragged-looking lobes. Taken once or twice upon trawled rock. Rare. Hastings.

60. *Sagartia rosea*, Gosse.

The tentacles of this species vary in colour from rose-red to crimson-lake or lilac. A most lovely anemone. It is usually found anchored down to some stone or mussel shell below the surface. Not very common. Hastings.

61. *Sagartia sphyrodeta*, Gosse.*
Hastings.

62. *Sagartia troglodytes*, Johnston.

This species occurs at Hastings in great variety, a favourite haunt being a mussel-bed with shingle beneath, the whole being covered with a thin layer of mud or sand. Here the anemones can attach themselves to the shingle or the mussel shells and withdraw instantly, or push their way upwards to expand on the surface. The species is nearly always known by the 'B' mark at the base of the tentacles, upon the inner face. Very common. Hastings.

63. *Sagartia viduata*, Müller.*
Hastings.

64. *Sagartia parasitica*, Couch.*
Brighton.

65. *Adamsia palliata*, Bohadsch.

Specimens of the form *rhodopis*, Gosse, have been taken upon shell of whelk and *Natica* from somewhat shallow water, and the variety *crinopis*, Gosse, upon shell of *Scaphander lignarius*. The *Acontia* are of a beautiful mauve colour and readily attract attention. Rare. Hastings.

ANTHEIDÆ

66. *Anthea cereus*, Ellis and Solander.

This beautiful species with low wide column and long, green, worm-like tentacles tipped with magenta occurs along the beach at Brighton. Upon the authority of Mr. Wells, of the Brighton Aquarium, the variety *rustica* is also a resident there. The species is certainly not known at Hastings, and the fact may indicate a difference of temperature of the water between these localities.

ACTINIIDÆ

68. *Actinia mesembryanthemum*, Ellis and Solander.

Characteristic of this species are the vivid blue dart-charged spherules around the margin of the disc, outside the tentacles. It occurs at Hastings in several varieties of colours, viz. vars. *a*, *β*, *ζ*, *ι* and *λ* of Gosse, the colours being respectively liver-brown, dark crimson, dark olive-green with broken lines of light green, and liver-coloured with green spots. Very common at low water. Hastings.

BUNODIDÆ

68. *Bunodes gemmacea*, Ellis and Solander.† Brighton.

69. *Bunodes clavata*, Thompson.† Brighton.

70. *Tealia crassicornis*, Müller.

A large and handsome anemone with wide and low column, the outer surface of which is provided with suckers. By these means the animal attaches to itself grains of sand and shell, covering itself to such an extent that it has often the appearance of a piece of stucco. The tentacles are short and thick, and generally barred with pink and white. The tentacles are occasionally found budding, the buds being produced from all sides. Common at low tide and from deeper water. Hastings.

ILYANTHIDÆ

71. *Ilyanthus mitchellii*, Gosse.

A rare species, and as such deserving fuller notice. A dozen specimens were obtained on one occasion from a trawler. Length of a specimen, 1½ inches. The colouring of the column varied as follows : In one instance it was wholly of an orange or light tomato-colour ; in others, and more generally, there was below the tentacles a flesh-coloured band, then a narrow or broad zone of tomato-colour extending to a quarter or half the length of the column, followed by a broad band of flesh-colour and another of tomato-colour of about equal depth, extending to the base. In one specimen the whole of the column was of a pale flesh tint, with the exception of two zones of a very pale tomato-shade.

The disc and tentacles were coloured as follows : Lip, opaque white, with an outer ring of brownish purple, then a wider zone of cream-colour and the space extending to the tentacles of brown-purple. The tentacles were in two rows ; the core of tentacle was of a light golden or straw colour, with bars upon the inner face of purple-brown, or in some of the outer tentacles of dark-grey ; the outer face of the tentacles appeared grey or curry-coloured. Around the base and upon each side of the tentacles swerved a cream-coloured line, not however uniting upon the outer side. Gonidial radii cream-coloured ; stomach a light tomato, with a line of deep orange-colour running down each ridge of the folds. The specimens were taken at the beginning of the year, and the white or salmon-coloured ova were clustered like grapes upon the mesenteries. Locality, 25 miles off Beachy Head.

ZOANTHIDÆ

72. *Zoanthus*, sp.

Upon scallop shells. Not uncommon. Hastings.

ALCYONARIA

ALCYONIDÆ

73. *Alcyonium digitatum*.

The only common coral upon the Sussex coast. It forms lobed, rounded masses upon rocks at low water and upon shells and rock from deeper water. The skeleton is spicular, and the polyps are white, with eight tentacles, fringed laterally with papillæ. The colour of the colonies is milk-white or orange. Common. Hastings.

CTENOPHORA

74. *Pleurobranchia pileus*.

Animal almost spherical, barely ½ inch in diameter, with eight longitudinal rows of swimming paddles, beneath each of which runs a circulatory canal terminating blindly at either extremity. The flash of the irridescent paddles in the sun as the little balloon-like body ascends in the water is a sight well worth seeing. Very common during most years, in the summer, at Hastings.

75. (?) *Pleurobranchia rhodopis*, Chun.

A large species of about the size and shape of a walnut, and of similar structure to the foregoing species. It was taken in the trawl in profusion a few years ago. The specimens were examined at the time, but for want of reference were left undetermined. Judging from memory however they corresponded precisely with Chun's figure of *P. rhodopis*, and cannot be referred to the balloon-shaped *Hormiphora plumosa*. Length, about 1¾ inches. Hastings.

76. (?) *Hormiphora plumosa*.

Specimens resembling this genus in shape and somewhat larger than *Pleurobranchia pileus* have been taken at Hastings, but were not examined sufficiently for identification. Hastings.

77. *Beroë ovata*, Eschscholtz.

This species is filbert-shaped, and has been taken in the immature condition. Specimen transparent, colourless, ¾ inch in length. In this genus the circulatory canals unite at the aboral pole, and in the region of the pores there are ciliated tentaculoid processes, lobed and branched. Hastings.

VERMES

CEPHYREA

1. *Sipunculus*, Pallasii.†
Brighton.

ANNELIDA

HIRUDINEA

2. *Pontobdella muricata*.

This species has suckers at both extremities ; there are neither feet nor bristles. The rings are strongly marked, and are studded the whole way round with conical warts, each ring somewhat resembling a well-armed dog collar. Colour, a dull buff with spots of brown at regular intervals. Trawled ; rare. Hastings.

POLYCHÆTA

AURICOMIDÆ

3. *Pectinaria belgica*.

This species forms beautifully made sand tubes, the sand grains being cemented together by an excretion, the structure resembling mosaic work. The tubes are straight, conical, and very regular. Common. Hastings.

4. *Pectinaria arenaria*.†
Brighton.

5. *Sabellaria crassissima*, Link.*
Very common. Hastings.

6. (?) *Sabellaria alveolata*.*
Common. Hastings.

7. *Siphonostoma*, sp.

Specimen about 2½ inches in length, greyish-brown in colour ; much enlarged towards the head. The setæ become very long in the region of the mouth and project in front. The body is covered with small papillæ. Trawled. Hastings.

TEREBELLIDÆ

8. *Terebella littoralis*, Dal.

This species, in the adult state, forms membranous tubes with agglutinated particles of sand and shell. The tubes are of considerable length and have a fringe of smaller branching tubes arranged in a radiating manner around the anterior end. Very common upon the shore. Hastings.

9. *Terebella maculata*.*
Hastings.

10. *Terebella conchilega*.†
Brighton.

SABELLIDÆ

11. *Sabella penicillus*.*
Hastings.

12. *Sabella tubularia*.*
Hastings.

13. *Myxicola infundibulum*.

Animal white, 1¼ inches long, tapering to the posterior end, and with a slight depressed line running down the dorsal and ventrical sides. Setæ minute. The pectinated plume-like gills are arranged around the mouth, and curve gracefully over towards the centre. Trawled ; rare. Hastings.

SERPULIDÆ

14. *Serpula contortuplicata*.

Annelid forming a calcareous tube, circular in section, much twisted and often aggregating together in involved masses. Animal with a flat-topped operculum. Very common upon shells and rock. Hastings.

15. *Serpula triquetra*.

This species inhabits a calcareous, serpentine tube cemented to shells and stone. Running down the back of the tube is a ridge or keel, and the base of attachment is spread out, so that a section would be somewhat triangular in form. The operculum is conical and generally furnished with two or three spines. Very common. Hastings.

MARINE ZOOLOGY

16. *Spirorbis nautiloides*, Link.

A minute species with coiled tube, the coils being arranged in the same plane. Attached to weed, corallines, and many other objects. Very common. Hastings.

17. *Spirorbis lucidus*, Montagu.*
Common. Hastings.

18. *Spirorbis communis.* †
Brighton.

19. *Filograna implexa.*

Animal secreting a long filiform, calcareous tube. The individuals are social, the associated tubes being packed together parallel-wise in considerable masses. Trawled ; not very common. Hastings.

ARENICOLIDÆ

20. *Arenicola piscatorum.*

The 'fisherman's worm' is about 6 inches long, brown, and with a large head which is covered with spiny tubercles. The dendriform gills are arranged along each side of the middle portion of the body. Very common upon the sand shore, where it burrows deeply. Hastings.

CIRRATULIDÆ

21. *Cirratula*, sp.
Hastings.

CHÆTOPTERIDÆ

22. *Chætopterus insignis*, Baird.

This species forms a tough parchment-like tube, 12 inches in length, by ¾ inch in diameter. The tubes are trawled upon the Diamond Ground, but so far as the writer has observed are always empty. Not uncommon. Hastings.

23. *Nereis margaritacea.*

Animal about 4 inches long, of a pearly lustre, and flattened dorsally and ventrally. It is nearly always found inhabiting the same shell with the trawled hermit crab ; this however is not the case with the smaller, shore-frequenting hermit crab. Common. Hastings.

24. *Nereis fimbriata*, Müller.*
Bexhill.

25. *Nereis brevimanus.* †
Brighton.

26. *Nereis longissima.* †
Brighton.

27. *Nereis bilineata.* †
Brighton.

28. *Nephthys*, sp.
Hastings.

29. *Phyllodoce viridis.* *
Common. Hastings.

AMPHINOMIDÆ

30. *Euphrosyne foliosa.*
Not uncommon. Hastings.

APHRODITIDÆ

31. *Aphrodite aculeata.*
Common. Hastings.

POLYZOA
ECTOPROCTA
CYMNOLÆMATA
CHEILOSTOMATA

ÆTEIDÆ

1. *Ætea anguina*, Linnæus.

A minute species with zoarium of ivory whiteness, usually found trailing over seaweed, and giving off from enlargements of the creeping stem or stolon, short tubular zoœcia, curved and expanded in the upper part, and with aperture terminal. The zoœcium bears some fancied resemblance to the arched head and neck of the hooded cobra. From moderately deep water. Not uncommon. Hastings.

2. *Ætea recta*, Hincks.

Resembling *Æ. anguina* but with zoœcium straight, and with longer aperture. Common upon rock and shells from the Diamond Ground and deep water. Hastings.

EUCRATIIDÆ

3. *Eucratea chelata*, Linnæus.

This dainty little species might well receive the popular appellation of the little glass slipper, for the zoœcia strikingly resemble a series of semi-transparent little slippers united together toe and heel by short prolongations. A free form, long, and rather tubular in the lower part of the zoœcium also occurs at Hastings, but is comparatively rare. Type, upon weed and coralline stems ; somewhat rare. Hastings.

4. *Gemellaria loricata*, Linnæus.

Colonies some 7 inches in height, seaweed-like and of a dull buff or grey colour. Stems formed of matted fibres, and the branches of zoœcia somewhat ovate, but tapering downwards, and arranged back to back. Not uncommon. Hastings.

CELLULARIIDÆ

5. *Scrupocellaria scruposa*, Linnæus.

Growing in little white branching tufts upon rock, etc. The branches consist of two rows of alternating zoœcia, both rows facing in the same direction. The zoœcia are ovately tubular, but taper downwards slightly, and the apertures are armed above with three or four spines. Upon the outer side of each zoœcium is a prominent avicularium with beak upturned. No operculum. From moderate to deep water. Common. Hastings.

6. *Scrupocellaria scrupea*, Busk.

Similar to the last species, but with a non-foliated operculum, and with a flagellate chamber situated between and rather behind the zoœcia. Upon sponges, etc. From moderate depth. Hastings.

7. *Scrupocellaria reptans*, Linnæus.

A species of prostrate, straggling habit, and of a dull buff or grey colour. Around the upper margin of the aperture there are three or four spines, and guarding the opening there is a foliated operculum. An avicularium is situated sometimes in front, and sometimes there is a smaller one behind the spines upon the outer side. Not uncommon upon *flustra*, etc. Hastings.

BICELLARIIDÆ

8. *Bicellaria ciliata*, Linnæus.

Colonies form plumose tufts of about 1 inch in height, of a dull grey colour, and hang pendent, like little tassels, upon the sides of the rocks, at low water. The stem and branches consist of biserial, alternating, glassy zoœcia, with many and long spines around the apertures, below and rather on the outer sides of which is situated a highly formed avicularium. The zoarium is a very beautiful object under the microscope. Common. Hastings.

9. *Bugula avicularia*, Linnæus.

Stem about 2 inches in height, giving off branches forming a delicate and beautiful spiral. The zoœcia are biserial and have a spine at the outer and upper corner, but only a very rudimentary one at the inner corner. The avicularia are longer than in *B. turbinata*, but not so prolonged as in Hincks' figures, and are placed midway down the aperture. Not uncommon upon rocks at low tide. Hastings.

10. *Bugula turbinata*, Alder.

Growth very much resembling the above species, but there are from two to five zoœcia abreast in a division, with a large spine at each of the upper corners of the aperture, and the aperture extends to the zoœcium below. The avicularia are shorter and wider than in *B. avicularia* and are placed just below the spines. Washed ashore. Hastings.

11. *Bugula flabellata*, Thompson.

A short, brown, truncate-looking growth. In some parts the zoœcia number seven in a row. There are two spines at each of the upper corners of the zoœcia, and often there is an avicularium half way down the side of the aperture. Common upon *Flustra foliacea*. Hastings.

12. *Bugula calathus*, Norman.

This species resembles very closely *B. flabellata*, but the habit is more compact and shorter. There are five or more zoœcia in a row, and in many cases three, but generally two spines in each of the upper corners. Hincks calls attention to the difference in colour between this species and the last. His observations hold good as regards the Hastings specimens, they being of a very pale buff colour. Washed ashore. Hastings.

13. *Bugula plumosa*, Pallas.*

Not uncommon. Hastings.

14. *Beania mirabilis*, Johnson.

A minute and delicate species creeping over rock, etc. The zoœcia are somewhat spoon-shaped, laterally compressed, with numerous spines around the margin, and connected together by prolongations from the preceding zoœcia. Common upon rock from the Diamond Ground. Hastings.

NOTAMIIDÆ

15. *Notamia bursaria*, Linnæus.

This exquisite little species throws up plume-like stems, much curled, and branching. The zoœcia are arranged in biserial and opposite order. Above each of these there projects from the stem a pedunculated avicularium. Upon weed; rather rare. Hastings.

CELLARIIDÆ

16. *Cellaria fistulosa*, Linnæus.

Zoœcia lozenge-shaped, arranged around a branching, jointed axis. The zoarium is white, and often occurs in dense bush-like masses. From moderate to deep water; rather common. Hastings.

17. *Cellaria sinuosa*, Hassal.

A much stouter species than the last, and consequently there are a greater number of zoœcia in the circumference. The upper margins of the zoœcia are curved, not straight as in *C. fistulosa*, and the segments of the stem are much longer. The colour is a light buff. From deep water ; rare. Hastings.

Flustridæ

18. *Flustra foliacea*, Linnæus.

Colonies forming long, flat, branching expansions, of a horny consistence, and with zoœcia arranged in lines and covering both surfaces. The zoœcia are coffin-shaped and carry two spines at either of the upper corners. The whole face of the zoœcium is membraneous. Very common upon shells, rock, etc. Hastings.

19. *Flustra papyracea*, Ellis and Solander.

This species occurs in the form of rather close tufts or rosettes about 2 inches in height. Zoœcia oblong, with only one spine at either upper corner. Colour buff ; not uncommon. Hastings.

Membraniporidæ

20. *Membranipora lacroixii*, Audouin.

Aperture of zoœcia oval, margin more or less beaded. It occurs upon rocks and stones at low tide, also upon shells, in three forms, viz. one producing considerable and uniform patches of stone ; secondly, it forms dendritic, and rather radiating patterns ; and, thirdly, there is a form with spines around the apertures, and producing colonies of a more or less close outline. Common. Hastings.

21. *Membranipora monostachys*, Busk.

Aperture of zoœcium oval, not occupying the whole width ; generally with one short and stout spine at the bottom, and often one or more on either side in the upper part ; occasionally there are none. The form of the colony is characteristic and might be expressed as erratically dendritic. Upon rock along the beach. Not very common. Hastings.

22. *Membranipora catenularia*, Jameson.

Zoœcia in single series, branches being given off at an open angle and uniting with others, thus forming reticulated patterns. The zoœcia are pear-shaped with oval and moderate sized apertures. Upon old shell of *Cardium norvegicum* from deep water. Rare. Hastings.

23. *Membranipora pilosa*, Linnæus.

Zoœcia glassy, perforated, with oval aperture occupying the full width of zoœcium, and armed with spines of which one at the bottom is very long and of a horny nature. Very common upon almost every object.

Membranipora pilosa var. *dentata*, having a short spine instead of a long one at the bottom of the aperture, is also common. Ranging from shore to deep water. Hastings.

24. *Membranipora membranacea*, Linnæus.

Covering rock and weed at low tide. Zoœcium coffin-shaped, brown, leathery, wrinkled, with a long spine upon the upper margin on either side, and occasionally one between them. Common. Hastings.

25. *Membranipora spinifera*, Johnston.* Hastings.

26. *Membranipora dumerillii*, Audouin.

Zoœcia form pearly patches upon rocks and stones at low water, also upon shells. They are oval to sub-triangular, with two spines on either side of aperture ; in some specimens the spines are abnormally long. Common. Hastings.

27. *Membranipora solidula*, Alder and Hincks.* Hastings.

28. *Membranipora aurita*, Hincks.

Forming patches upon stone at low tide. Zoœcium ovately oblong ; margin finely beaded, with a spine upon one side only, below the aperture. Not uncommon. Hastings.

29. *Membranipora flemingii*, Busk.

Zoœcium ovate, aperture sub-triangular and occupying rather more than half the front area, the other portion forming a calcareous wall. There are three spines upon either side of the upper half of the aperture. Upon scallop shells ; rather rare. Hastings.

30. *Membranipora rosselii*, Audouin.

Zoœcium coffin-shaped, margin strongly beaded, aperture sub-triangular and occupying barely half the length of the zoœcium ; colonies forming patches upon rock and shells. From moderately deep water ; rather rare. Hastings.

31. *Membranipora savartii*, Audouin.

Forming considerable patches upon old shells of oyster, *Lutraria*, etc. The zoœcia are arranged in lines and arched above and below; the front is closed in to the extent of one-fourth or one-third, leaving an aperture oval to sub-rectangular, with a few irregular projections around the edge. There is a raised and crenulated margin to the zoœcium, and the space between this and the aperture is also crenulated in the upper part. New to Britain. From deep water; rather rare. Hastings.

MICROPORIDÆ

32. *Micropora coriacia*, Esper.

Forming small rounded patches upon old shells. Zoœcia coffin-shaped; front wall finely perforated; margin with a tubercle at each corner of the aperture; aperture semi-circular. Upon *Pecten opercularis* from deep water; somewhat rare. Hastings.

CRIBRILINIDÆ

33. *Cribrilina radiata*, Moll.

Occurring in various forms, encrusting old shells. Zoœcium oval, aperture semi-circular and usually armed with five spines. The front wall is radiately ribbed, the extremities of the ribs often developing into blunt erect spines. Beneath the aperture there is generally more or less of a boss or umbo. Upon oyster and other shells from rather deep water; common. Hastings.

34. *Cribrilina punctata*, Hassall.

This species also occurs in several forms, producing patches upon old shells. The front wall is radiately punctured, and the aperture is semicircular with two spines upon the upper margin, the lower margin being thickened. In another form there are five spines to the upper margin and an avicularium at each corner. Not uncommon upon oyster, scallop shells, etc., from rather deep water. Hastings.

35. *Cribrilina figularis*, Johnston.

Zoœcia forming strongly marked patches upon old shells. They are ovate with sub-square apertures, the lower margin having a shallow sinus. Upon the front wall there is an inner oval area radiately punctured, the outer margin being plain. Avicularia very large. From rather deep water; somewhat rare. Hastings.

36. *Membraniporella nitida*, Johnston.

Zoœcia oval, of crystalline brightness, forming round patches. Aperture nearly semicircular; front wall formed of radiating ribs. Occurring upon *Lepralia foliacea* from deep water; rare. Hastings.

37. *Membraniporella melolontha*, Busk.

Forming glistening, more or less foliated or branching patches upon old shells. The species resembles *nitida*, but there is a spine at either corner of the aperture and a plain margin around the ribbed area. Quite characteristic however is the spinous process at the bottom of this area, and very suggestive of a little tail. Upon old oyster and other shells; rather rare. Hastings.

MICROPORELLIDÆ

38. *Microporella ciliata*, Pallas.

Forming nacreous patches upon shells, weed, etc. Zoœcium ovate; aperture with five spines, sometimes six, around the upper margin, and with a raised pore beneath the lower margin. Upon one side there is usually an avicularium with a mandible of extreme length, although not as yet of the vibraculoid type. From moderately deep water; very common. Hastings.

39. *Microporella malusii*, Audouin.

A very handsome species. Zoœcia ovately lozenge-shaped; aperture semi-circular with three spines upon the upper margin. Beneath the lower margin there is a crescentic pore, and smaller dimpled pores occur over the front wall. This species forms white patches upon old shells, stones, etc. From moderate to deep water; common. Hastings.

40. *Microporella impressa*, Audouin.

Zoœcia coffin-shaped, rather long, the lower half pointed; aperture somewhat more deeply arched than a semicircle. Beneath it there is situated a pore, other minor ones being distributed over the front wall, more particularly around the margin. Colonies have a beautiful satin-like lustre. Upon weed from moderately deep water; not uncommon. Hastings.

41. *Microporella violacea*, Johnston.

Colonies rather extensive, of a violet lustre, encrusting old shells; the zoœcia are coffin-shaped, aperture semicircular to ob-ovate. In the middle of the front wall is a depression with a characteristic pore in the centre. Around the margin are radial vacancies unfilled in with shelly matter. Very common upon shells and stones from deep water. Hastings.

42. *Chorizopora brongniarti*, Audouin.

Forming delicate pearly patches upon old shells, stones, etc. Zoœcia ovate but tapering rather below, transversely wrinkled, with aperture approximately semi-circular and a more or less pronounced prominence underneath. The connection with adjacent zoœcia is not continuous, but vacant spaces occur at intervals around the margin. Common upon scallop shells, etc., from deep water. Hastings.

PORINIDÆ

* 43. *Porina tubulosa*, Norman.

Zoœcia ovate in the lower half, the upper part narrowed and curving upwards, with a circular aperture. A little below this there is a tubular pore, and elsewhere the wall is pitted with pores. Upon old shell of *Cardium norvegicum*; rare. Hastings.

44. *Lagenipora socialis*, Hincks.*

Hastings (Miss Jelly).

MYRIOZOIDÆ

45. *Schizoporella unicornis*, Johnston.

Encrusts rock at low tide; zoœcia rather large, square to sexagonal; aperture semi-circular with sinus in the lower lip. Beneath there is usually a short blunt spine, and in either or both upper corners an avicularium. In one instance noted, an avicularian zoöid had usurped the position of an ordinary zoöid, the aperture of the latter appearing as a minute pore immediately above the mandibular apparatus of the former, the zoœcium remaining of the normal size but with partial obliteration of outline. Common. Hastings.

46. *Schizoporella vulgaris*, Moll.*

Hastings.

47. *Schizoporella simplex*, Johnston.

Encrusting old shells. Zoœcia ovate; aperture elevated and with a sinus to the primitive orifice, the matured one being circular. Beneath the aperture, and generally confluent with it, is a blunt prominence. Oœcia with a few irregular spiny protuberances. Not very common; from moderate to deep water. Hastings.

48. *Schizoporella linearis*, Hassall.

Encrusting old shells. Zoœcia oblong, arranged in lines; aperture round, with a small sinus in the lower margin; front wall with pores. Upon one or both sides of the aperture, and a little below, is placed an avicularium pointing towards it. Common. Hastings.

49. *Schizoporella bi-aperta*, Michelin.

Zoœcia more or less oblong; aperture round with a sharp sinus below; front wall plain. Upon one or both sides of the aperture there is a considerable prominence surmounted by an avicularium. Upon *Pecten opercularis*, etc. From moderately deep water; rather rare. Hastings.

50. *Schizoporella auriculata*, Hassall.

Forming round patches upon shells, stones, etc. Zoœcia square to oblong; primary aperture round with a sharp sinus in the lower margin; secondary margin forming a wide loop below, enclosing a short tubular pore (? avicularium). Beneath the aperture is a strong prominence. Oœcium sometimes crescentic, sometimes orbicular. Not uncommon. Hastings.

51. *Schizoporella discoidea*, Busk.

Zoœcia angular, rather short, and with the front wall slightly pitted. Primary aperture has five spines and a narrow sinus; mature orifice circular and raised, but not with an angular or pointed lip as shown in Hincks' figures. Just below, and to the right or left of the aperture, there is a small round avicularium. The oœcia greatly overlap the zoœcia. Rather rare. Hastings.

52. *Schizoporella hyalina*, Linnæus.

Encrusting seaweed. Zoœcia with a satin-like gloss, ovate but tapering and curved below. The aperture is round and has a small sinus in the lower lip, beneath which there is a slight umbo. Not uncommon. Hastings.

53. *Schizoporella venusta*, Norman.

Upon dead shells. Zoœcia glistening, lozenge-shaped to sexagonal; aperture sub-ovate, slightly disjunct, below which there is a prominence. From moderate to deep water; rather rare. Hastings.

54. *Mastigophora hyndmanni*, Johnston.

Zoœcium ovate; aperture sub-circular with a narrow sinus; upon either side is situated a considerably modified avicularium, of which the chamber outline is often preserved. The mandible is greatly elongated, even more so than in *Microporella ciliata*, and has now more the appearance of a tapering stick or whip, and is termed by Hincks a vibraculum. Encrusting an old scallop shell from deep water; rare. Hastings.

55. *Schizotheca fissa*, Busk.

Zoœcia ovate, small, short ; aperture elevated, with six spines and a narrow sinus in the lower lip ; oœcium sub-crescentic or with a wedge-shaped fissure in the middle. Upon an old scallop shell from deep water ; rare. Hastings.

56. *Hippothoa divaricata*, Lamouroux.

Colonies composed of oval zoœcia arranged in single sequence and connected together by tubular prolongations. Lateral branches are given off at an open angle and unite with others ; the aperture is circular but with a narrow sinus below ; beneath the aperture runs a thickened, longitudinal, median ridge. Common upon shells and stones from moderately deep water. Hastings.

57. *Hippothoa flagellum*, Manzoni.

Similar in habit to the last species, but the zoœcia are farther apart, the interval being usually equal in length to two zoœcia, whereas in *divaricata* it is usually equal to the length of one. The aperture is egg-shaped, and there is no median ridge. Common. Hastings.

58. *Rhyncopora bi-spinosa*, Johnston.

A species encrusting old shells, and one subject to several modifications in the region of the aperture, the margin of which is sometimes produced into two lateral and vertical processes ; at others, one of these may be bent across the aperture, or may become central. A fairly constant feature is a more or less spinous mucro rising from beneath the aperture. There is often too a large avicularium mounted upon a broad pedestal, taking the place of this mucro, and generally placed rather laterally. One specimen obtained is found enveloping a colony of the hydroid *Hydractinia echinata*, itself encrusting the shell of a *Nassa*. The oœcia upon this specimen are particularly plentiful. Trawled ; rather rare. Hastings.

ESCHARIDÆ

59. *Lepralia pallasiana*, Moll.

A hardy looking species encrusting rocks at low tide. Aperture large, more deeply arched than a semicircle ; front wall pitted. Common. Hastings.

60. *Lepralia foliacea*, Ellis and Solander.

This species forms large masses of foliated and anastomosing laminæ, the zoœcia being disposed on both sides of the laminæ. The zoœcia are ovate, and have large pores over the front wall. Aperture horse-shoe-shaped, with sometimes a slight prominence beneath. Habitat, rather deep water. Somewhat scarce. Hastings.

61. *Lepralia pertusa*, Esper.

Encrusting old shells. Zoœcia oval, with a circular aperture, of which the lower margin is slightly disjunct ; beneath is a process, tri-radiate in form ; wall of zoœcium poriferous. It may be noted that young zoœcia of *Smittia cheilostoma* before the development of the sinus closely resemble the zoœcia of this species. From moderately deep water; rather rare. Hastings.

62. *Lepralia adpressa*, Busk.

Colonies encrusting small shells, e.g. whelk, *Natica*, *Trochus*. The zoœcia are ovate, pitted, and have a boss at either or both corners of the aperture. The latter is horseshoe shaped, the sides being slightly indented ; from moderately deep water ; somewhat rare. Hastings.

63. *Lepralia nitidula*, Hincks, MS.

Hastings.

64. *Porella concinna*, Busk.

A species encrusting shells and stones and showing much variation. The zoœcia are coffin-shaped, or various in form, and has an opalescent lustre. The margin is often deeply sinuous, almost dove-tailed, and is perforated along the border. Aperture horseshoe shaped, with two spines in marginal zoœcia and a prominence below. Very common. From moderate to deep water. Hastings.

65. *Smittia lanasborovii*, Johnston.

Zoœcia crystalline, oblong or coffin-shaped, with perforations in the front wall. Aperture nearly round and raised into a collar, with a small avicularium upon the lower lip. Not uncommon upon rock from deep water. Hastings.

66. *Smittia reticulata*, McGillivray.

In this species there is a deep sinus in the lower lip, and instead of an avicularium there, as in the last species, there is a larger one below, pointing downwards and looking like a pendant from a neck. Above the aperture there are either two or three spines, and the margin of the zoœcia is bordered with pits or vacancies. Hincks remarks upon an instance where two zoœcia side by side have a single wide oœcium in common. The writer has observed nearly the same phenomenon, and in a specimen before him the union of two zoœcia is seen in different phases of completeness. The cause is evidently overcrowding, the zoœcia in which it occurs being extremely narrow. Associated with the last species ; not uncommon. Hastings.

67. Smittia cheilostoma, Manzoni.

Occurring as light red patches upon old shells and stones. Zoœcia coffin-shaped, rather pointed below ; front wall perforated ; aperture raised, sub-circular, but with a large sinus in the lower margin ; within is seen a flat-topped process. As already stated elsewhere, the immature zoœcia of this species resemble nearly the zoœcia of *Lepralia pertusa*. Common from deep water. Hastings.

68. Smittia trispinosa, Johnston.

Forming large buff-coloured patches upon old shells, stones, etc. Aperture raised, sharp, with a sinus in the lower lip, and in some cases four spines above. Below the aperture on one side, and pointing towards it, is an avicularium ; and around the margin of the zoœcium are a series of pits or perforations. From moderate to deep water ; very common. Hastings.

69. Phylactella labrosa, Busk.

Upon shells. Zoœcia ovate, short, perforated, and arranged in single divergent lines. The aperture is round, raised and expanded, and shows a small denticle within. From deep water ; somewhat rare. Hastings.

70. Phylactella collaris, Norman.

Colonies more compact than in the last species. The front wall is plain, and there is no denticle upon the lower lip of the aperture, which is sometimes slightly pointed. Upon old shells. From moderate to deep water ; not very common. Hastings.

71. Mucronella peachii, Johnston.

Encrusting dead shells, stones, etc. Zoœcia ovate to lozenge-shaped ; aperture round, with six spines around the upper margin ; upon the lower lip, within, is a double-pointed denticle, and upon the outer lip a small conical tooth. The area below the aperture is rather swollen. Common from moderate to deep water. Hastings.

72. Mucronella ventricosa, Hassall.

This species much resembles the last, but comparison, by the aid of a handglass only, will show colonies to be of coarser grain, or composed of larger zoœcia. These are ventricose, the aperture has four instead of two spines, and there is a tongue-like process projecting from immediately below the lower lip. Common upon shells from deep water. Hastings.

73. Mucronella variolosa, Johnston.

Colonies encrusting dead shells. When in good condition they have a strong violet lustre. The zoœcia are coffin-shaped and pitted around the margin ; aperture, round to subquadrate, with two long spines upon the upper rim, and a small process or tooth upon the lower one ; behind which is seen a flat-topped process. Common from moderate to deep water. Hastings.

74. Mucronella coccinea.

A very handsome species of a violet lustre. The zoœcia are rather short, wide, and with marbled markings. The aperture in marginal zoœcia is circular, with six spines upon the upper rim and a pointed process upon the lower lip. Upon both the right and left side of the aperture is a formidable-looking avicularium, one of them sometimes being very large. Not uncommon upon rock and other objects from deep water. Hastings.

Mucronella coccinea var. *mammilata*, answering exactly to Hincks' description, occurs also at Hastings. Upon shell of *Pectunculus glycymeris*.

75. Palmicellaria skenei, Ellis and Solander.*
Hastings.

CELLEPORIDÆ

76. Cellepora pumicosa, Linnæus.

This species by successive layers of zoœcia forms rounded masses of a few inches in diameter, upon scallop and other shells. The zoœcia are oval, upright, with circular aperture, and a long, pointed rostrum arising from beneath it, and carrying a small avicularium. Common from deep water. Hastings.

77. Cellepora avicularis, Hincks.*
Hastings.

78. Cellepora costazii, Audouin.*
Hastings.

Cellepora costazii var. *tubulosa*, Hincks.*
Hastings.

79. *Cellepora ramulosa*, Linnæus.

Partially encases the stems of corallines. The zoœcia are ovate to tubular, with a stout prominence beneath the aperture but not so vertical as in *C. pumicosa*, and with a rather large avicularium at the base, and upon the inner surface. Very common from the coralline zone. Hastings.

CYCLOSTOMATA

CRISIIDÆ

80. *Crisia cornuta*, Linnæus.

Colonies forming erect, feather-like growths upon seaweed and corallines. The zoœcia are tubular and curved, giving off other zoœcia from behind in single series. Near the base of each zoœcium there is a horny joint ; and in many cases at the side, and a little way below the aperture, there occurs a long, tapering, curved spine also having horny joints. Common from moderately deep water. Hastings.

A variety without spines, but not of Hincks' *geniculata* type, also occurs somewhat rarely at Hastings.

81. *Crisia eburnea*, Linnæus.

Of similar habit to the last species. The zoœcia are biserial and alternate with only the ends free. Horny joints occur at intervals along the stem and branches, but always at the commencement of each branch. The oœcia are pear-shaped, and occupy the position of a zoœcium.

Crisia eburnea var. *aculeata*, Hassall, with a long jointed spine upon the off-side of the aperture, also occurs, together with the type, at Hastings, both being rather common.

82. *Crisia denticulata*, Lamarck.

Much like the last species, but the zoœcia are more compact, not so elongated, apertures not so distant, more opposite, and the space between the two lines of zoœcia is greater ; the habit moreover is straighter. Not very common. Hastings.

TUBULIPORIDÆ

83. *Stomatopora granulata*, Milne-Edwards.* Hastings.

84. *Stomatopora major*, Johnson.

A species forming little straggling, irregularly-branching colonies upon rock, shells, etc. The zoœcia are tubular and number in the widest part as many as seven abreast, the number increasing with the length of the branch. The anterior ends of the zoœcia curve upwards and are free, showing a circular aperture. This occurs either irregularly or, as is often the case, in rows. Not uncommon from moderately deep water. Hastings.

85. *Entolophora clavata*, Busk.* Hastings.

86. *Tubulipora lobulata*, Hassall.

Encrusting shells, and forming somewhat radiatingly lobed or branching colonies, of a mauve colour. The tubular zoœcia are enlarged in the upper part, but contract toward the aperture which is upturned and free. The primary tubes multiply rapidly, producing fan-like expansion of the lobes. From deep water ; somewhat rare. Hastings.

87. *Tubulipora flabellaris*, Fabricius.

Forming thin, flat, fan-like colonies upon scallop and other shells. Apertures slightly enlarged and raised. The oœcia are seen as oval expansions occupying the width of three or four zoœcia. From deep water ; common. Hastings.

88. *Idmonea serpens*, Linnæus.

Exquisite, mauve or purple coloured colonies, upon corallines, flustra, etc. The upturned anterior ends of the tubular zoœcia occur mainly in rows, upon either side of a central parting. Colonies branch by dividing. Not uncommon from moderately deep water. Hastings.

89. *Diastopora patina*, Lamarck.

Colonies forming little white discs of radiating tubular zoœcia, upon stone, corallines, etc. Occasionally colonies take a concave form. Around a small area in the centre the anterior ends of the zoœcia are erect and free, but outside remain horizontal and do not rise above the common matrix. Around the edge of the colony is seen the white border of a fine calcareous carpet, spread around for the due reception of the dainty polypides. Rather common from moderate to deep water. Hastings.

90. *Diastopora obelia*, Johnston.

Encrusting shells, etc. The zoœcia radiate from a centre, the apertures being barely elevated above the colonial crust. Lines forming the boundary between the adjacent rows of zoœcia, sinuous and distinctly marked. Dotted here and there between the apertures are small tubular orifices, reproducing in miniature the zoœcial apertures. Very common from deep water. Hastings.

MARINE ZOOLOGY

91. *Diastopora sarniensis*, Norman.

Forming encrusting patches upon shells and rocks. The zoœcia radiate from a centre, the anterior ends curving upwards and being free. The apertures are elliptical and alternate, and in many cases closed, except as regards a small tubular orifice which projects from the operculum. There is a plain white basal border surrounding the colony. Common from deep water. Hastings.

92. *Diastopora sub-orbicularis*.

Colonies forming circular crusts upon dead shells. The zoœcia radiate from a centre, the apertures being elliptical, alternate, and only occasionally raised. Zoœcia and opercula minutely perforated. From deep water ; not very common. Hastings.

LICHENOPORIDÆ

93. *Lichenopora hispida*, Fleming.

This species forms little mounds consisting of tubular zoœcia arranged radially, the free ends projecting beyond the common matrix ; the apertures are thrown into several sharp points. Colonies also occur in compound form of greater extent, the surface appearing dimpled. There is a wide border of the basal layer displayed around the colonies. Very common upon shells from deep water. Hastings.

CTENOSTOMATA

ALCYONIDIIDÆ

94. *Alcyonidium gelatinosum*, Linnæus.

Colonies consisting of zoöids embedded in irregularly lobed masses of a gelatinous matrix. Specimens from deep water are often very large and intricately lobed. Colonies are smooth and of a light buff or brown-green colour. Common. Hastings.

95. *Alcyonidium hirsutum*, Fleming.

In occasional years this species is found covering weed at low tide in profuse masses. The surface of colonies is mammilated and the colour a buff-brown. Hastings.

96. *Alcyonidium parasiticum*, Fleming.

Forming inconsiderable colonies upon coralline stems ; colour, grey-brown. When the zoöids are withdrawn the surface is thrown somewhat into wrinkles. From the coralline zone ; not common. Hastings.

97. *Alcyonidium mytili*, Dalyell.*
Hastings.

FLUSTRELLIDÆ

98. *Flustrella hispida*, Fabricius.*
Somewhat rare. Hastings.

VESICULARIIDÆ

99. *Vesicularia spinosa*, Linnæus.

This species forms erect, horny, branching growths of fine texture, and when dry resembles in colour and appearance so many strands of tow. The main stems are bent in zig-zag manner, giving off branches at each bend, but they are concealed by a number of finer, climbing tubes which also branch. The zoœcia are ovately cylindrical and are arranged uniserially and equidistantly upon the branches. Not uncommon from moderate to deep water. Hastings.

100. *Amathea londigera*, Linnæus.

Forms little intricate masses, consisting of horny, branching stems bearing at regular intervals linear groups of about eight zoœcia, resembling in imagination so many little sacks stacked together. The stems repeatedly divide, the division taking place immediately after each group of zoœcia Upon weed, etc., from moderately deep water. Not uncommon. Hastings.

101. *Bowerbankia imbricata*, Adams.†
Brighton.

102. *Bowerbankia pustulosa*, Ellis and Solander.

Little shrub-like growths of about 1½ inches in height. Zoarium horny, brown in the lower parts, and branching at an angle of about 30°. The zoœcia are ovately cylindrical and occur in biserial and slightly spiral order at the end of each branch, or immediately before re-branching takes place. Upon trawled rock ; rare. Hastings.

103. *Farrella repens*, Farre.*
Hastings.

Farrella repens var. *elongata*.*
Hastings.

BUSKIIDÆ

104. *Buskia nitens*, Alder.*
Hastings.

CYLINDRŒCIIDÆ

105. *Anguinella palmata*, Van Beneden.

Occurring in pendent, mud-coloured clusters upon rocks at low tide. Length from 3 to 4 inches. The stems, with their short and palmate branches, are rather catkin-like. The zoœcia are long and tubular, and do not differ in appearance from the axis. Colonies in texture are rather india-rubber-like. Common. Hastings.

VALKERIIDÆ

106. *Valkeria uva*, Linnæus.

Favourite habitat, over-running *Corallina officinalis*. Zoœcia ovately cylindrical, slightly narrowing towards the aperture. Rather common. Hastings.

Valkeria uva var. *cuscuta.**
Hastings.

ENTOPROCTA

PEDICELLINEA

PEDICELLINIDÆ

107. *Pedicellina cernua*, Pallas.

Species over-running *Corallina officinalis*, throwing up short pedicels, each bearing a polypide and a zoœcium at the top. The pedicel is spinous, and as in the other members of this genus, flexible, which is demonstrated by the zoöid making a motion as of bowing or nodding. In this species the pedicel tapers slightly towards the top, but is not constricted at the apex, as in *P. nutans*; and further the zoöid is greatly more protuberant upon the anal than upon the oral side.

Pedicellina cernua var. *glabra*, having a smooth pedicel, also occurs at Hastings together with the type. Both are common.

108. *Pedicellina nutans*, Dalyell.*
Hastings.

109. *Pedicellina gracilis*, Sars.*
Hastings.

ECHINODERMA

HOLOTHUROIDEA

1. *Synapta inhærens*, O. F. Müller.

Specimens wormlike, of a pale flesh colour; about 2 inches in length and ⅛ inch in diameter. The body has no podia; the tentacles are twelve in number and bilaterally lobed. Spicules occur in the form of anchors and perforated plates; the latter are egg-shaped in outline, and both are devoid of serrations. On some occasions the fishermen's nets are choked with these creatures. Rare generally. Hastings.

2. *Cucumaria pentactes*, Forbes.†
Brighton.

3. *Cucumaria lactea*, Forbes and Goodsir.

Animal about 1¼ inches in length, of a light chocolate-brown colour. Podia alternate, in five rows; discs of suckers, white. There are ten dendriform tentacles which, together with the disc, are of a light buff colour freckled with brown, the tentacles becoming quite pale towards the tips. The spicules are nodulated, perforated plates. A single specimen trawled half a mile from shore. Rare. Hastings.

4. *Thyone fusus*, O. F. Müller.

Specimens about 3 inches in length, flesh-coloured to pink; test rather delicate. The podia are numerous and scattered generally over the body, but in some cases show a tendency to longitudinal arrangement. The tentacles are dendroidal, and ten in number, two of them being smaller than the others and having red cores; these two tentacles are constantly applied to the mouth. The tentacles and disc are powdered with brown over the pink ground, and the mouth is of a dark brown. The spicules are sub-rectangular tables with two-legged central pieces. From the Diamond Ground; rare. Hastings.

5. *Thyone fusus* (?), O. F. Müller.

Specimens white, barely ½ inch in length, probably immature. The podia are plentiful and scattered, displaying however some longitudinal arrangement. The spicules are perforated tables with two-legged central pieces, the immature ones being somewhat lozenge-shaped, and the mature ones ovate to sub-quadrangular. There is some little variation between the tables of these specimens and those of the last species, but they are probably referable to the same species. Associated with scallops. Shoreham.

6. *Phyllophorus drummondi*, Thompson.

Specimens white, about 5 inches in length, tapering below; test rather tough. The podia are scattered rather thinly, and occur to some extent along longitudinal lines. Tentacles seventeen in number, dendroidal, alternating in size; stems of tentacles rather brown, mouth of a dark brown. Spicular tables rather large, circular to sub-quadrangular in outline. From the Diamond Ground; rare. Hastings.

7. *Holothuria nigra*, Kinahan.†
Brighton.

ASTEROIDEA

ASTERINIDÆ

8. *Palmipes placenta*, Penn.

Species with five arms connected by a

web. Not uncommon in the trawl. Hastings.

SOLASTERIDÆ

9. *Solaster papposus*, Fabricius.

The 'sun' starfish, usually having thirteen rays and covered with spinous papillæ. Colour, purplish red. Not uncommon in the trawl. Hastings.

ECHINASTERIDÆ

10. *Henricia sanguinolenta*, O. F. Müller.*
Hastings.

ASTERIIDÆ

11. *Asterias rubens*, Linnæus.

The common five-fingered starfish. Very common. Hastings.

12. *Asterias hispida*, Penn.†
Brighton.

13. *Asterias aurantiaca*.†
Brighton.

OPHIUROIDEA

OPHIOLEPIDIDÆ

14. *Ophiura ciliaris*, Linnæus.

This species has five arms, smooth and snake-like, and is of a grey-buff colour. Common in the trawl. Hastings.

15. *Ophiura albida*, Forbes.†
Brighton.

OPHIOTHRYCIDÆ

16. *Ophiothrix fragilis*, Abilg.

Species with five very spiny arms, and one in which the variation in colouring is absolutely infinite. The most usual colours are white, salmon, pink, red, sage-green, grey, brown, etc. When handled alive they readily detach fragments of the arms, so that it is almost impossible to secure a complete specimen. When however the fishermen's nets are hung up to dry and the brittle stars are allowed to dry unhandled they may be taken in perfect condition, the arms displaying every possible curve and contortion. They are known by the fishermen as 'castle cats.' Not a shore species. Hastings.

AMPHIURIDÆ

17. *Amphiura elegans*, Leach.

A small species, common upon the rocks at low water. Hastings.

18. *Ophiocnida brachiata*, Montague.

A species with five very long arms. Rare. Hastings.

ECHINOIDEA

ECHINIDÆ

19. *Echinus esculentus*, Linnæus.
(?) Hastings.*   Brighton.†

20. *Echinus miliaris*, Gmel.

Test circular, mouth central, anus apical; colour purple. Very common in the trawl and sometimes met with at extreme low tide. Hastings.

21. *Strongylocentrotus lividus*, Lamarck.†
Brighton.

22. *Strongylocentrotus dræbachiensis*, O. F. Müller.†
Brighton.

CLYPEASTRIDÆ

23. *Echinocyamus pusillus*, O. F. Müller.

A small depressed, heart-shaped species, measuring about ¼ inch in length; colour, green; mouth, central; vent, midway between the mouth and margin. Trawled; not uncommon. Hastings.

SPATANGIDÆ

24. *Spatangus purpureus*, O. F. Müller.

A large heart-shaped, purple species. The mouth is situated midway between the centre and the margin; the spines upon the under surface are a favourite habitat of the minute bivalve mollusc *Montacuta substriata*. Trawled in rather deep water; somewhat rare. Hastings.

25. *Echinocardium cordatum*, Penn.

Species heart-shaped, with rather fine spines; colour, grey; test, mouse-like; mouth midway between the centre and the margin. Rather common in the trawl, and occasionally cast on shore. Hastings.

BRACHIOPODA
INARTICULATA

CRANIIDÆ

1. *Crania anomala*.†
Brighton.

MOLLUSCA
AMPHINEURA
POLYPLACOPHORA

CHITONIDÆ

1. *Tonicella ruber*, Lowe.†
Brighton.

2. *Callochiton lævis*, Montague.†
Brighton.

3. *Craspedochilus onyx*, Spengler.
Trawled upon rock from rather deep water ; common. Hastings.

4. *Craspedochilus albus*, Linnæus.†
Brighton.

5. *Craspedochilus cinereus*, Linnæus.
Upon rocks near low water ; not uncommon. Hastings.

6. *Acanthochiets fascicularis*, Linnæus.
Common upon rocks at low tide. Hastings.

PELECYPODA
PROTOBRANCHIA
NUCULIDÆ

7. *Nucula nucleus*, Linnæus.
Common from the coralline zone. Hastings.

8. *Nucula nitida*, Sowerby.*
Somewhat rare. Hastings.

9. *Nuculana minuta* var. *brevirostris*, Jeffreys.
Rare. Rye Bay.

FILIBRANCHIA
ANOMIACEA
ANOMIIDÆ

10. *Anomia ephippium*, Linnæus.
Not uncommon upon trawled rock, etc. Hastings.

Anomia ephippium var. *aculeata*, Müller.
Small ; rare. Hastings.

11. *Anomia patelliformis*, Linnæus.
Often within or upon other dead bivalve shells. Not uncommon ; trawled. Hastings.

ARCACEA
ARCIDÆ

12. *Glycymeris glycymeris*, Linnæus.
Common upon the Diamond Ground. Hastings.

Glycymeris glycymeris var. *pilosa*, Linnæus.
Common. Hastings.

Glycymeris glycymeris (?) var. *globosa*, Jeffreys.
Hastings.

13. *Barbatia lactea*, Linnæus.
Rather rare. Hastings.

MYTILACEA
MYTILIDÆ

14. *Mytilus edulis*, Linnæus.
Very common. Hastings.

Mytilus edulis var. *pellucida*, Pennant.
Somewhat rare. Hastings.

15. *Volsella modiola*, Linnæus.
Not uncommon. Hastings.

16. *Volsella barbata*, Linnæus.
Upon trawled rock and shells ; common. Hastings.

17. *Volsella adriatica*, Lamarck.*
Very rare. Hastings.

18. *Modiolaria marmorata*, Forbes.
Harboured within the tests of *Tunicates*, and amongst the root fibres of the hydroid *Antennularia*. Not very common. Hastings.

19. *Modiolaria discors*, Linnæus.†
Brighton.

PSEUDOLAMELLIBRANCHIA
OSTREIDÆ

20. *Ostrea edulis*, Linnæus.
Common. Hastings.

PECTINIDÆ

21. *Pecten maximus*, Linnæus.
The scallops from the English side of the Channel are much covered with animal growth ; those from the French side are much cleaner and more variegated in colour. The winter of 1895–6 was so severe that the cold killed off all the scallops from the Hastings grounds, and the beds have not as yet been replenished, only one or two being occasionally taken. Hastings.

22. *Hinnites pusio*, Linnæus.
Upon trawled rock, etc. Rather rare. Hastings.

23. *Chlamys varius*, Linnæus.
A shell running through many most delicate shades of yellow, orange, puce and brown. Moored by the byssus to rocks, dead shells, etc. ; trawled. Hastings.

24. *Æquipecten opercularis*, Linnæus.
Shell displaying every shade of colour between white, yellow, orange, brown and purple, with combinations of these colours.

MARINE ZOOLOGY

Æquipecten opercularis var. *lineata*, da Costa.
Shell with lines of a darker colour than the ground running down each rib. Both forms common. Hastings.

LIMIDÆ

25. *Lima subauriculata*, Montague.*
Hastings.

26. *Lima loscombi*, Sowerby.†
Brighton.

27. *Lima hians*, Gmelin.†
Brighton.

EULAMELLIBRANCHIA
SUBMYTILACEA

CYPRINIDÆ

28. *Cyprina islandica*, Linnæus.
Rather rare. Hastings.

LUCINIDÆ

29. *Lucina borealis*, Linnæus.
Rather rare. Hastings.

30. *Montacuta substriata*, Montague.
Habitat upon the spines of the underside of the sea-urchin, *Spatangus purpureus*. From the Diamond Ground, off Hastings.

LEPTONIDÆ

31. *Kellia suborbicularis*, Montague.*
Rare. Hastings.

TELLINACEA

SCROBICULARIIDÆ

32. *Syndosmya prismatica*, Montague.
Rather rare. Hastings.

33. *Syndosmya alba*, Wood.
Common. Hastings.

34. *Scrobicularia plana*, da Costa.
Occurring at the mouth of the Rother and in the channels leading into it. Common. Rye Harbour.

TELLINIDÆ

35. *Tellina crassa*, Gmelin.
From the Diamond Ground. Not uncommon. Hastings.

36. *Tellina tenuis*, da Costa.
A delicate little shell of various colours : white, yellow, pink, buff. Common. Hastings.

37. *Tellina fabula*, Gronovius.
Shell in appearance somewhat like that of the last species, but not so large, and the right valve is lined with diagonal striations. Common. Hastings.

38. *Macoma balthica*, Linnæus.
A small, strong shell ; colour, white, yellow, or red. Common. Hastings.

DONACIDÆ

39. *Donax vittatus*, da Costa.
At low water this species may be detected in large communities by the little mounds of sand which they cast up in digging into the sand with the foot. Very common. Hastings.

MACTRIDÆ

40. *Mactra stultorum*, Linnæus.
Very common upon the sand shore. Hastings.

Mactra stultorum var. *cinerea*, Montague.
Not uncommon. Hastings.

41. *Spisula solida*, Linnæus.
Rare. Hastings.

42. *Spisula elliptica*, Brown.
Trawled ; somewhat rare. Hastings.

43. *Spisula subtruncata*, da Costa.
Shell triangular, thick in the umbonal region ; rather rare. Hastings.

44. *Lutraria elliptica*, Lamarck.
Trawled upon the Diamond Ground ; not uncommon. Hastings.

45. *Lutraria oblonga*, Chemnitz.
Single valves only taken, and those in a very deteriorated condition. Trawled ; rather rare. Hastings.

VENERACEA

VENERIDÆ

46. *Lucinopsis undata*, Pennant.
Trawled ; rather rare. Hastings.

47. *Dosinia exoleta*, Linnæus.*
Rare. Hastings.

48. *Dosinia lupina* (*lincta*), Linnæus.
Rather rare. Hastings.

49. *Venus fasciata*, da Costa.
Brighton.

50. *Venus casina*, Linnæus.
Shell pale in colour ; ornamented with concentric ribs or lamellæ ; rare. Hastings.

51. *Venus verrucosa*, Linnæus.
Ornamentation resembling that of the last species, but the ribs are less pronounced except in front and behind, where they are interrupted and form folds. Not uncommon. Hastings.

52. *Venus ovata*, Pennant.

Shell small, radiately ribbed, and with fine concentric lines ; rather rare. Hastings.

53. *Venus gallina*, Linnæus.*

Very rare. Hastings.

54. *Tapes virgineus*, Linnæus.

Shell laterally compressed, polished. Colour more or less in rays, undulatory. Trawled ; common. Hastings.

55. *Tapes pullastra*, Montague.

Very common. Hastings.

Tapes pullastra var. *perforans*, Montague.

Shell rather smaller than the type ; not so deep from umbo to margin ; common. Hastings.

56. *Tapes decussatus*, Linnæus.

Shell somewhat quadrangular, strongly cancellated with radial and concentric lines ; somewhat scarce. Hastings.

CARDIACEA

CARDIIDÆ

57. *Cardium echinatum*, Linnæus.

Common. Hastings.

58. *Cardium exiguum*, Gmelin.*

Very rare. Hastings.

59. *Cardium nodosum*, Turton.

A single valve only ; rare. Hastings.

60. *Cardium edule*, Linnæus.

Common. Rye.

61. *Cardium norvegicum*, Spengler.

Common in the trawl. Hastings.

MYACEA

GARIDÆ

62. *Gari ferrænsis* (Chemnitz).

Brighton.

MYIDÆ

63. *Mya truncata*, Linnæus.

A coarse shell, mainly conspicuous by the membranous, siphonal sheath ; common. Hastings.

64. *Mya arenaria*, Linnæus.

A single valve only taken, and that of an immature specimen ; rare. Hastings.

65. *Sphenia binghami*, Turton.*

Rare. Hastings.

SOLENIDÆ

66. *Solecurtus scopula*, Turton, var. *oblonga*, Jeffreys.

Shell proportionately not so deep from

umbo to margin as in the type. Trawled about 25 miles off Beachy Head ; rare. Hastings.

67. *Cultellus pellucidus*, Pennant.†

Brighton.

68. *Ensis ensis*, Linnæus.

Shell curved, small ; very common. Hastings.

69. *Ensis siliqua*, Linnæus.

Shell nearly straight ; somewhat common. Hastings.

70. *Solen vagina*, Linnæus.

Shell short and straight.

SAXICAVIDÆ

71. *Saxicava rugosa*, Linnæus.

This little ugly deformed shell is found boring into the hardest rock. Not uncommon. Hastings.

GASTROCHÆNIDÆ

72. *Gastrochæna dubia*, Pennant.†

Brighton.

PHOLADACEA

PHOLADIDÆ

73. *Pholas dactylus*, Linnæus.

Shell large, delicate and white. Animal with boring propensities; common. Hastings.

74. *Barnea candida*, Linnæus.

Common. Hastings.

75. *Barnea parva*, Pennant.*

Somewhat rare. Hastings.

76. *Zirfæa crispata*, Linnæus.

Shell short and deep ; not uncommon. Hastings.

TEREDINIDÆ

77. *Teredo navalis*, Linnæus.†

Brighton.

ANATINACEA

PANDORIDÆ

78. *Pandora inæquivalvis*, Linnæus.

Shell with the left valve flat, and the right one convex ; rather rare. Hastings.

Pandora inæquivalvis var. *pinna*, Montague.

Shell not so produced in front as in the type ; rather rare. Hastings.

LYONSIIDÆ

79. *Lyonsia norvegica* (Chemnitz)†

Brighton.

ANATINIDÆ

80. *Thracia fragilis*, Pennant.
Rare. Rye Harbour.

Thracia fragilis var. *villosiuscula*, MacGillivray.
Brighton.†

SEPTIBRANCHIA

CUSPIDARIIDÆ

81. *Cuspidaria cuspidata* (Olivi) †
Brighton.

SCAPHOPODA

DENTALIIDÆ

82. *Dentalium vulgare*, da Costa.
Very common. Hastings.

83. *Dentalium entalis*, Linnæus.†
Brighton.

GASTROPODA
PROSOBRANCHIA

ASPIDOBRANCHIA

PATELLIDÆ

84. *Patella vulgata*, Linnæus.
The limpet ; very common. Hastings.

Patella vulgata var. *depressa*.
Brighton.†

85. *Helcion pellucidum*, Linnæus.
A specimen shows alternate rays of blue and brown upon a dull orange ground ; rare. Hastings.

ACMÆIDÆ

86. *Acmæa virginea* (Müller)
Shell pink, with interrupted lines of colour ; rare. Hastings.

FISSURELLIDÆ

87. *Emarginula fissura*, Linnæus.
Shell depressed, with a marginal slit in front ; rare. Hastings.

88. *Emarginula conica*, Schumacher.
Shell more conical than that of the last species ; not uncommon upon trawled rock. Hastings.

89. *Fissurella græca*, Linnæus.
The 'key-hole' limpet. Shell cancellated with radial and concentric ribs, and having an apical perforation ; somewhat rare. Hastings.

TROCHIDÆ

90. *Gibbula magus*, Linnæus.
Shell depressed, with large umbilicus ; common. Hastings.

91. *Gibbula tumida*, Montague.*
Rare. Hastings.

92. *Gibbula cineraria*, Linnæus.
Shell slightly turreted ; common. Hastings.

93. *Gibbula umbilicata*, Montague.
Shell conical, sides slightly convex ; common. Hastings.

94. *Calliostoma montagui*, Wood.
Shell rather high in the spire, with spiral ridges, and fine striæ crossing the intervening furrows ; rare. Hastings.

95. *Calliostoma exasperatum*, Pennant.†
Brighton.

96. *Calliostoma granulatum* (Born)
Shell acutely conical, granulated ; sides slightly concave ; not uncommon. Hastings.

Calliostoma granulatum var. *lactea*, Jeffreys.
Similar to the type, but white ; rather rare. Hastings.

97. *Calliostoma zizyphinus*, Linnæus.
Shell acutely conical, sides straight, whorls spirally lined ; common. Hastings.

Calliostoma zizyphinus var. *lyonsi*, Leach.
Shell white and nearly smooth ; rather uncommon. Hastings.

TURBINIDÆ

98. *Phasianella pulla*.
Shell small, conical, and with lines of red running diagonally across the whorls ; very rare. Hastings.

PECTINIBRANCHIA

LITTORINIDÆ

99. *Lacuna crassior*, Montague.
Periostracum horny-looking, wrinkled concentrically ; rare. Hastings.

100. *Lacuna divaricata* (Fabricius)
Brighton.

101. *Lacuna parva* (da Costa)
Brighton.

102. *Lacuna pallidula* (da Costa)
Brighton.

103. *Littorina obtusata*, Linnæus.
Shell with low spire. Colour, canary or orange ; common. Hastings.

104. *Littorina littorea*, Linnæus.
The edible periwinkle ; very common. Hastings.

105. *Littorina rudis* (Maton)†
Brighton.

Littorina rudis var. *saxatilis*, Johnston.
Brighton.†

Littorina rudis var. *tenebrosa*, Montague.
Brighton.†

106. '*Littorina canaliculatus.*' †
(Recorded thus) Brighton.

107. '*Littorina lichina.*' †
(Recorded thus) Brighton.

RISSOIIDÆ

108. *Rissoa parva*, da Costa.*
Hastings.

Rissoia parva var. *interrupta*, Adams.
Hastings.

109. *Rissoia inconspicua* var. *ventrosa*,
Jeffreys.†
Brighton.

110. *Rissoia violacea*, Desmarest.†
Brighton.

111. *Alvania lactea* (Michaud)
Hastings.

112. *Manzonia costata* (J. Adams) †
Brighton.

113. *Zippora membranacea* var. *labiosa*,
Montague †
Brighton.

114. *Onoba striata* (J. Adams) *
Hastings.

115. *Cingula semistriata*, Montague.*
Hastings.

PALUDESTRINIDÆ

116. *Paludestrina stagnalis* (Basterot)
Shell conical ; whorls smooth and nearly
flat ; common. Rye.

Paludestrina stagnalis var. *octona*, Lin-
næus.
Whorls rather convex ; common.
Hastings.

TRUNCATELLIDÆ

117. *Truncatella truncata* (Montague) †
This species is recorded as *T. montagui.*
Brighton.

CYPRÆIDÆ

118. *Trivia europæa*, Montague.*
The only British species ; common.
Hastings.

NATICIDÆ

119. *Natica catena* (da Costa)
Shell rather globular, whorls convex,
suture distinct ; common. Hastings.

120. *Natica alderi*, Forbes.
Shell half the size of the last species,
very glossy ; whorls not so convex, suture
not so deep ; common. Hastings.

LAMELLARIIDÆ

121. *Velutina lævigata*, Pennant.
Spire depressed ; body-whorl large ; rare.
Rye Harbour.

122. *Velutella flexilis*, Montague.†
Brighton.

CERITHIIDÆ

123. *Bittium reticulatum* (da Costa) *
Rare. Hastings.

124. *Cerithiopsis tubercularis*, Montague.*
Hastings.

SCALIDÆ

125. *Scala clathrus*, Linnæus.
Shell spiral ; whorls furnished with concentric ribs and mauve coloured spiral lines ;
somewhat rare. Hastings.

126. *Scala clathratula*, Adams.*
Rare. Hastings.

127. *Aclis minor*, Brown.*
Very rare. Hastings.

PYRAMIDELLIDÆ

128. *Brachystomia ambigua* (Maton &
Rackett) *
Hastings.

129. *Pyrgulina interstincta* (?) (Montague) *
Hastings.

130. *Pyrgulina clathrata*, Jeffreys.
Brighton.

131. *Spiralinella spiralis*, Montague *
Hastings.

132. *Turbonilla lactea* (Linnæus) *
Hastings.

EULIMIDÆ

133. *Eulima polita*, Linnæus.
Shell white, polished, with long spire
and flattened whorls ; rare. Hastings.

TURRITELLIDÆ

134. *Turritella communis*, Lamarck.*
Hastings.

APORRHAIDÆ

135. *Aporrhais pes-pelecani*, Linnæus.*
Hastings.

BUCCINIDÆ

136. *Buccinum undatum*, Linnæus.
The whelk ; common. Hastings.
The left-handed monstrosity, the keeled monstrosity and the acuminate form * also occur at Hastings, but are rare.

137. *Donovania minima*, Montague.*
Hastings.

138. *Neptunea antiqua*, Linnæus.*
This species is recorded as ' *Fusus antiquus*,' Linnæus, and may be received with a little caution, since the animal is often imported into the town with the common whelk, as food, the shells being thrown upon the beach. Hastings.

139. *Tritonofusus gracilis* (Costa)
Shell with the spire not quite in a straight line ; somewhat rare. Hastings.

MURICIDÆ

140. *Ocinebra erinacea*, Linnæus.
Common. Hastings.

141. *Ocinebra corallina*, Scacchi.†
Brighton.

142. *Trophon clathratus*, Linnæus.†
(?) British species. Brighton.

143. *Purpura lapillus*, Linnæus.
This species occurs in great variety of colouring, mainly white, yellow, orange or brown. Very common. Hastings.

NASSIDÆ

144. *Nassa reticulata*, Linnæus.
Both this species and the following one are frequenters of the lobster ' pots.' Very common. Hastings.

145. *Nassa incrassata* (Ström)
Very common. Hastings.

146. *Nassa pygmæa*, Lamarck.*
Rare. Hastings.

PLEUROTOMIDÆ

147. *Bela turricula*, Montague.
Common in the trawl, dead shells only having been taken. Hastings.

Bela turricula var. *rosea*, M. Sars.
Rare. Hastings.

148. *Bela rufa*, Montague.*
Somewhat rare. Hastings.

149. *Mangilia costata* (Donovan)
Very rare. Hastings.

150. *Mangilia rugulosa* (Philippi)
Rare. Hastings.

151. *Mangilia nebula* (Montague)†
Brighton.

152. *Clathurella linearis*, Montague.*
Very rare. Hastings.

153. *Clathurella reticulata*, Renier.*
Very rare. Hastings.

OPISTHOBRANCHIA
TECTIBRANCHIA

ACTÆONIDÆ

154. *Actæon tornatilis*, Linnæus.
Rare. Hastings.

SCAPHANDRIDÆ

155. *Scaphander lignarius*, Linnæus.
Rare. Hastings.

156. *Bullinella cylindracea* (Pennant) †
Brighton.

PHILINIDÆ

157. *Philine aperta*, Linnæus.
Rather local ; plentiful in Rye Bay. Hastings.

CAVOLINIIDÆ

158. *Cavolinia trispinosa*, Rang.
Very rare. Hastings.

NUDIBRANCHIA

HERMÆIDÆ

159. *Hermæa dendritica*.
Species a little over ¼ inch in length, the general colour varying from seaweed green to orange. There is only one pair of tentacles, and they are characteristic of the genus, being folded lengthwise. The papillæ are rather long and overhang the sides and middle of the back, giving to the animal a shaggy appearance. Further characteristic of the species is the dentritic veining, of a deeper colour, which runs down each side of the back supplying offshoots to the tentacles and papillæ, the veins being more or less reticulated. The animal has a habit of contracting the head and tail, and in that condition much resembles a young specimen of the anemone *Actinia mesembryanthemum*, the papillæ representing the tentacles of the anemone. This species is decidedly rare. It occurred at Hastings during a rather hot summer, upon rocks at low spring tides, but has not been observed since. It is very difficult to detect. Hastings.

EOLIDIDÆ

160. *Eolis papillosa*, Linnæus.

A fine species, and one that is always in summer to be found on our coast. Length, about 2 inches. The body is flesh-coloured, powdered with opaque white ; the papillæ, which well clothe the sides, are somewhat lanceolate in form and are usually of a seaweed green finely speckled with opaque white. The colours however run through many gradations, as is seen in years of great abundance, when all shades of grey, green, orange, brown and brown-purple make their appearance. Habitat, along the shore under stones and crawling upon mud and sand. Common. Hastings.

161. *Æolidella alderi*, (?) Cocks (? *glauca*).

A single specimen taken upon *Lepralia foliacea* from moderately deep water appears almost intermediate in character between this species and *Æolidella glauca*. In form perhaps the specimen approaches nearer to *Æ. alderi*, but in colouring, size and habitat approximates to *Æ. glauca*. Length of specimen 1 inch, extending when crawling to $1\frac{1}{4}$ inches. The papillæ are neither 'vermicular' nor 'clavate,' but are thickest at about the middle, slightly conical at the apex, where there is in many cases a small pimple-like inflation of the outer membrane. Under magnification the internal gland is of a seaweed green-brown, the sheath and apex being pellucid white superficially and indefinedly banded with opaque sage-green paling to white, with a glow of orange upon the upper portions. The papillæ of the front row are semi-transparent white. The oral tentacles are a little longer than the dorsal ones, and are lightly tinged with orange ; the latter are linear and obtuse at the apex and of a bright orange colour tipped with white. The tail is short and not very pointed. From the veil, and extending midway down the back, the colour is a bright orange, paling towards the middle of the back. Hastings.

162. *Cuthona aurantiaca*, Alder and Hancock.

Animal $\frac{1}{2}$ inch in length. The papillæ are rather long ; gland of a rosy orange colour finely granulated with red. The apex and sheath are semi-opaque blue-white ; below the apex and internally is a zone of opaque white granules ; above this zone and overlapping it somewhat is a surface colouring of orange, toning down to yellow. Of the oral tentacles, the sheath is transparent, colourless, and the core semi-opaque white, rather defined. This species bears some resemblance to the following species, but may be distinguished by the absence of foot processes, and by the presence of an orange-coloured zone near the apex of the papillæ. Upon water-logged timber ; trawled ; rare. Hastings.

163. *Cratena concinna*, Alder and Hancock.

Length of specimen $\frac{1}{2}$ inch. The papillæ are often carried bent ; they have a sheath of blue-green or bottle-green, the gland being orange coloured, or brownish-orange speckled with dark brown ; but very characteristic is a crown around the apex, formed of linear opaque white granules, arranged perpendicularly to the surface. These are generally very strongly marked, but are occasionally absent, or nearly so. The foot has two blunt processes. Rare. Hastings.

164. *Tergipes despectus*, Johnston.

This little species is barely $\frac{1}{4}$ inch in length, and when met with cannot be confounded with any other. The body is semi-transparent white, with a tinge of red in front of and behind the dorsal tentacles. Upon either side of the back and arising from a central vessel are three or four club-shaped papillæ, with glands of a mottled seaweed-green colour, and apices of opaque white. The only habitat is upon shore hydroids, particularly *Obelia gelatinosa* and *O. geniculata*, where it will probably be first detected by the little globular masses containing the eggs which are deposited upon the stems. It may be noted that the papillæ of this little creature, when gliding over the stems of the hydroid, bear a striking resemblance to the egg capsules of the latter, and hence may reap some immunity from danger. During some years this species is very plentiful. Hastings.

165. *Galvina cingulata*, Alder and Hancock.

Animal $\frac{1}{2}$ inch in length, long and slender and with very tapering tail. The character of the papillæ in this species has very faithfully suggested a comparison to the quills of the 'fretful porcupine.' In the Hastings specimens there are eight rows of papillæ, each papilla having three surface bands of a marbled seaweed-green colour ; above these is one of opaque white.

The gland is foliated and of an opaque whitish yellow, and somewhat granular. The papillæ when observed were slightly contracted, and in that condition, which appeared to be more or less normal, they became slightly constricted at each band, thus giving an undulating outline to the papillæ ; the reason for this lay in the fact that the internal gland was narrower at those points. The dorsal tentacles are almost twice the length of the oral ones ; they are long and linear and with obtuse tips, the latter being transparent white with a band of opaque white below, and about midway down there is a patch of seaweed-green colour. In 1900 about half a dozen specimens were taken upon *Plumularia pinnata*, to which had been attached about a dozen of their egg-bands. In the next year two more specimens, together with *Doto coronata*, were taken at low water, upon *Obelia longissima*. In the former instance the resemblance of the papillæ to the egg capsules of the hydroid was rather striking. Rare. Hastings.

166. *Galvina tricolor*, Forbes.

Length of specimen 1 inch. General appearance : animal of a pale orange or buff colour ; papillæ inflated and pointed, those in the front half or two-thirds of the body being tipped with orange ; the hinder papillæ also being orange-tipped, and in addition speckled with dark brown. The dorsal tentacles are colourless and have obscure internal lines of opaque white running down each, to bulbous bases and the eyes. Tail colourless. Closer examination of a papilla shows a constriction near the end, the end portion being divided into three zones of colour ; that nearest the constriction is opaque white with brown-black spots, the next is of orange or pale yellow, and the one at the apex colourless and transparent ; the lower portion of the papilla is semi-opaque white, a deeply lobed gland of a pale straw colour being occasionally seen, which sends a single straight stem-like lobe to the apex. A single specimen trawled from moderately deep water. Rare. Hastings.

167. *Coryphella rufibranchialis*, Johnston.
  Eolis pellucida, Alder and Hancock.
  Eolis gracilis, Alder and Hancock.

Length of animal, 1¼ inches ; body semi-transparent white. The papillæ are rather long and linear, the apex is pellucid white, below which is nearly always a ring of opaque white, of a granular character, and internal. The sheath is transparent colourless, and the gland of a bright orange-red or coral colour, uneven in outline, squared at the top and leaving only a small margin of the sheath showing. The dorsal tentacles are wrinkled and have a faint tinge of orange, the tips being granulated with opaque white ; they are a little longer than the oral ones, which are also tipped with opaque white granules upon the inner and upper side ; the foot processes are long. In March 1897 an abnormally coloured specimen was obtained from the coralline zone. In this specimen the veil, both pairs of tentacles, and the upper side of the end of the tail were coloured a beautiful mauve violet, excepting the apices of the tentacles which were opaque white. Some of the papillæ also partook somewhat of the mauve colour. Since the above date another specimen taken upon the shore (not its usual habitat) showed the same tendency towards mauve colouring. The deep-water form is much longer, and has papillæ more filamentous than the variety from shallower water. Habitat from moderately shallow water to the coralline zone. Occasionally plentiful. Hastings.

168. *Coryphella landsburgi*, Alder and Hancock.

The length of this beautiful little creature is not much more than $\frac{1}{3}$ inch. Animal in part colourless, semi-transparent, but coloured along the sides of the body, the papillæ and their bases, and over the head and hinder part of the body, with light violet or mauve. Looked at more closely, the central gland is orange, the sheath of a light violet, and there is a band or patch of opaque white near the apex. Both pairs of tentacles are violet for one-third of the way up them, the upper part being transparent white, with a slight patch of opaque white at the apex ; foot processes moderately long. Trawled half a mile from shore upon flustra, where it was only detected by its colour. Rare. Hastings.

169. *Facelina coronata*, Forbes and Goodsir.

In summer this gorgeously coloured species is always more or less with us. The length is 1¼ inches ; the body is pellucid grey slightly speckled with blue. In the papillæ, which are tapering, the core varies in colour from light to dark orange or red-brown ; the sheath is pellucid grey, with an irregular splash of very vivid blue about the centre, and a ring of opaque white around the apex, with spots of opaque white scattered below. The dorsal tentacles are ringed, and are of a pale buff or orange colour, the oral ones being long and sweeping. Habitat, under stones and upon rocks at low tide. Common. Hastings.

170. *Facelina drummondi*, Thompson.

Animal ⅔ inch in length ; body translucent, of a faint rose colour ; back, pale salmon. The dorsal tentacles have from twenty to thirty laminæ, are blunt at the apex which is of a pale orange tint, the tentacle itself being rose-orange and having in front a narrow patch of granular, opalescent white, extending one third of the way down ; there are also spots of a similar character between and in front of the tentacles. The oral tentacles are a trifle longer and more pointed than the dorsal ones, and somewhat similarly coloured. As regards papillæ, notes are unfortunately deficient. Obtained from shrimp trawler. Rare. Hastings.

171. *Antiopa cristata*, delle Chiaje.

An immature specimen barely ¼ inch in length. Body transparent, colourless, or with a slight tinge of buff. The papillæ are inflated, or club-shaped with pointed tips, the foremost three upon either side projecting greatly in advance of the tentacles and appearing to act somewhat as tactile organs, contracting at every contact with any object ; their tips form a broad line in advance of the animal ; the hinder papillæ are large and project considerably beyond the tail. The gland of the papillæ is very narrow and linear, enlarging or dividing near the apex and coloured purplish-brown or greenish-brown, the outer portion being transparent, but of the same hue. The apex is opaque white and where this meets the colour below, the result is a metallic blue. The dorsal tentacles are laminated and are short and blunt, with an opaque white spot at the apex ; between them is a raised and warty frontal piece, coloured buff. A single specimen taken upon weed in a rock pool at low water. Rare. Hastings.

DOTONIDÆ

172. *Doto coronata*, Gmelin.

Animal a little over ⅓ inch in length, of a pale orange colour, except as regards the foot which is colourless. There is a single pair of tentacles having characteristic trumpet-mouthed sheaths. Along either side of the body are arranged eight club-shaped papillæ, each one being encircled with about eight equidistant rings of large tubercles, each tubercle having at its apex a distinct black spot ; internally there are opaque white egg-like bodies.

This species is common upon the coralline *Antennularia* to which it attaches its egg-band ; also taken upon *Obelia longissima*. Hastings.

173. *Doto pinnatifida*, Montague.

Animal very similar to the last species, but with a more or less interrupted band down the back of brownish-green ; and along each side of the cloak are from one to three rows of tubercles set alternately, each tubercle containing an opaque white body and in some cases a black speck also. Similar tubercules with like contents occur irregularly down the back, the black specks being conspicuous. These black specks are of peculiar interest since it has been surmised that they may be organs of vision. They also occur around the rim of the tentacle-sheath and on lobular processes arising from the latter. Common upon *Antennularia* from the Diamond Ground. Hastings.

DENDRONOTIDÆ

174. *Dendronotus frondosus*, Ascanius.

Animal mottled light and dark brown, or buff, sometimes almost golden ; in some cases the colour is uniform. Along either side of the back are from six to eight branching, tree-like gills. The tentacles are ringed and arise from a sheath the rim of which also gives off branching processes. Specimens obtained only a little distance from shore are small, that is from 1 to 2 inches in length, but some trawled upon the Diamond Ground must have measured close upon 5 inches. Rather rare. Hastings.

TRITONIIDÆ

175. *Tritonia hombergi*, Cuvier.

Animal fully 3 inches in length ; back warty, and slate coloured. Around the margin of the cloak are arranged rather closely a number of dendriform gills. The tentacles have sheaths and near the apex branched processes are given off. This creature is not prepossessing, being in the preserved condition rather toad-like. Not uncommon from the Diamond Ground. Hastings.

DORIDIDÆ

176. *Archidoris tuberculata*, Cuvier.*

Rather rare ; upon rocks at low water.
Hastings.

177. *Jorunna johnstoni*, Alder and Hancock.

A large orange coloured species, nearly
smooth. Taken two or three times upon
rocks at low spring tides. Rather rare.
Hastings.

POLYCERIDÆ

178. *Ægirus punctilucens*, d'Orbigny.†

Brighton.

179. *Palio lessoni*, d'Orbigny.

A single specimen obtained of this
species agrees with Alder and Hancock's
description in having short tentacles with
ten or twelve laminæ, but the body is
more ocellated, and the frill across the snout
is as in var. *ocellata*. The body is more
tuberculated perhaps than in the variety and
has the line of tubercles along the head, but
there are others also. The animal has the
habit of spinning a thread of mucus at the
tail and thereby mooring itself to some
object ; it is also fond of screwing up the
tail end of the foot into a regular disc, and
suspending itself by it, from the surface of
the water.

Palio lessoni, var. *ocellata*, Alder and
Hancock.

Animal of a seaweed-green colour,
covered with white or greenish white spots
and tubercles. There is a single pair of
tentacles which are laminated and of a
green colour, in the upper part. The gills
consist of three branching plumes facing
backwards, and anal aperture being in the
centre. Running along each side of the
body from the front of the head, and
terminating behind the gills in free processes is a pretty little frill. From this
point, a line of white spiny tubercles runs
down the centre of the tail, and similar
tubercles occur generally distributed over
the upper portion of the body. This
variety during some years is rather plentiful
upon rocks and under stones at low water,
but in others it appears to be practically
absent. Hastings.

180. *Polycera quadrilineata*, Müller.

This species has much the character of
the foregoing species. The body is transparent white, with orange spots and spiny
tubercles of the same colour down the
centre of the back and tail, in the latter

part of which they merge to form a streak
of orange. Another line of tubercles runs
down each side of the body commencing in
front of the head, where they form four
rather long points or processes projecting
forward, and running backwards as far as
the side of the gills where they terminate
as free processes. There is a single pair of
tentacles which are laminated in the upper
part. The gills are plume-like and form an
incomplete rosette situated about midway
down the back. Taken a few times in the
trawl a mile or so distant from shore.
Rather rare. Hastings.

181. *Acanthodoris pilosa*, Müller.

Animal nearly 1 inch in length, of a pale
yellow or white colour. Cloak covered
with rather long and conical tubercles.
Not uncommon upon rocks from mid to
low tide. Hastings.

182. *Lamellidoris bilamellata*, Linnæus.

The cloak is of a pale, or rich brown
colour, and is tuberculated, the tubercles
being white. The branchiæ form a double
rosette of plumes in the form of the letter
Omega inverted ; tentacles laminated.
Very common upon rocks from mid to low
tide, often congregating in groups. Hastings.

183. *Lamellidoris diaphana*, Alder and
Hancock.†

Brighton.

184. *Goniodoris nodosa*, Montague.

Animal ¾ inch in length, of a pale
yellow colour, excepting the back which is
flesh-coloured. There are two oral tentacles, and the dorsal ones are laminated
behind in the upper half. The free margin
of the cloak is very prettily frilled and
turned up and may or may not unite behind
the gills which form a rosette of plumes at
the end of the back. The whole upper
portion of the animal is more or less dotted
with opaque white, and down the centre of
the back there is a keel, and upon either
side a few small tubercles are arranged in
line and coloured opaque white ; down the
centre of the tail there is an opaque saffron-
coloured ridge. This is a mud-loving
species and is therefore rather local. Hastings.

185. *Ancula cristata*, Alder.

Animal fully ½ inch in length ; transparent white. The gills which occupy the
centre of the body are formed of three
branching plumes which face backwards.

Upon either side of these and arising from the body are four or five erect, linear processes. The dorsal tentacles are laminated in the upper half, that portion being bent backwards. From near the base of each tentacle are given off two narrow arms directed forwards ; and upon either side of the lip there is a tentacular process. All of these appendages are more or less tipped with orange or yellow, besides which there is a line down the tail of the same colour. This species is very beautiful when under full sail in its proper element, but when seen upon the rocks at low tide it appears like a little shapeless ball of jelly quite unlikely to attract attention. It is usually found in moderate profusion. Hastings.

PULMONATA

AURICULIDÆ

186. *Alexia myosotis* (Draparnand)
Somewhat rare. Rye.

CEPHALOPODA

DIRBRANCHIATA

DECAPODA

OMMASTREPHIDÆ

187. *Todarodes sagittatus*, Lamarck.*
Very rare. Hastings.

LOLIGINIDÆ

188. *Loligo vulgaris*, Lamarck.
The 'Squid.' Common. Hastings.

SEPIIDÆ

189. *Sepia officinalis*, Linnæus.
The cuttle-fish. Common. Hastings.

SEPIOLIDÆ

190. *Sepiola scandica*, Steenstrup.
Common. Hastings.

OCTOPODA

POLYPODIDÆ

191. *Polypus vulgaris*, Lamarck.
Not uncommon during unusually warm summers. Hastings.

192. *Moschites cirrosa*, Lamarck.
Upon the authority of Mr. W. Wells of the Brighton Aquarium this species occurs off that town. It is peculiar for having a single row of suckers along each arm.

CHORDATA

UROCHORDATA

ASCIDIACEA

ASCIDEÆ SIMPLICES

ASCIDIADÆ

1. *Ascidia mentula*, O. F. Müller.

Test semi-opaque, rather coarse, of a faint yellowish flesh colour ; height, about 2 inches. Branchial aperture with eight lobes, and both apertures coloured bright crimson or with bright red spots. The test often has embedded in its exterior coralline stems, polyzoa, etc., and internally harbours the mollusc *Modiolaria marmorata*. Common in the trawl and sometimes cast ashore. Hastings.

2. *Ascidia virginea*, O. F. Müller.

Test transparent, smooth, colourless ; height 1¼ inches. Specimens often enfolded in the branches of corallines, the test sometimes extending a short way up the stems. Branchial aperture eight-lobed. Often harbours *Modiolaria marmorata*. Not uncommon ; trawled. Hastings.

3. *Ascidia vitrea*.†

Brighton.

4. *Ascidia*, sp.

Upon the shore at extreme low water, are often found extensive beds of an Ascidia, packed closely together. The test is coated with sand and is oblong to subglobular in form ; length of a specimen about ¾ inch. Branchial orifice appears to be four-lobed, but the lobes are probably subdivided ; about four short tentacles are seen. Common. Hastings.

5. *Ascidia*, sp.

There is a species, somewhat resembling the last mentioned, occurring upon corallines. It is smaller, measuring about ¼ inch in diameter, rather depressed, and coated with sand. Lobes of the branchial orifice as in the foregoing species. Common. Hastings.

6. (?) *Molgula arenosa*, Alder and Hancock.

Animal nearly globular, barely ¾ inch in diameter ; coated with fine sand ; branchial orifice six-lobed. Specimen has been apparently slightly attached. Tubular character of orifice non-apparent. Hastings.

7. *Cynthia morus*, Forbes.

Animal barely ½ inch in width, wider than high, warty, attached by a wide base ; colour, magenta crimson ; branchial orifice with four pale lobes. A specimen preserved for some years in formalin has successfully retained its colour. Somewhat rare ; trawled. Hastings.

ASCIDIÆ COMPOSITÆ

BOTRYLLIDÆ

8. *Botryllus schlosseri.*†
Brighton.

9. *Botryllus polycyclus.*†
Brighton.

10. *Botryllus bivittatus.*†
Brighton.

11. *Leptoclinum listerianum.*†
Brighton.

12. *Distoma rubrum.*†
Brighton.

13. *Amaroecium*, sp.

Specimen consists of a little flat gelatinous mass, ¼ inch in thickness, and 1½ inches in length, lobed around the edge, and with the margin folded over upwards and inwards, like a clenched fist. One or two delicate algæ pass up through the centre. The mass is translucent, colourless, and the zoöids are of an opaque grey-buff, and are seen here and there in well defined oblong systems. In other portions the systems become indistinct, and in parts, there is no arrangement at all. There is no common aperture. The zoöid has a six-lobed branchial orifice, a long atrial languet, a long post-abdomen, which terminates in a flattened rounded lobe, with another lobe more or less well formed on either side. Anus, from lateral view, is bilobed ? Embedded in the gelatinous matrix is a little sand. The specimen does not appear to correspond with either of the species *proliferum*, *nordmanni*, or *argus* but bears some resemblance in the lobed character of the specimen to *Aplidium fallax*. In combining the two characters of systemic and non-systemic arrangement of the zoöids the specimen bears a resemblance to *A. irregulare* var. *concinnum*, Herdman, from different latitudes. Rare. Hastings.

14. *Amaroecium argus.*†
Brighton.

www.ingramcontent.com/pod-product-compliance
Lightning Source LLC
Chambersburg PA
CBHW021719210326
41599CB00013B/1697